New Agricultural Crops

AAAS Selected Symposia Series

New Agricultural Crops

Edited by Gary A. Ritchie

CRC Press
Taylor & Francis Group
Boca Raton London New York

CRC Press is an imprint of the
Taylor & Francis Group, an **informa** business

First published 1979 by Westview Press

Published 2018 by CRC Press
Taylor & Francis Group
6000 Broken Sound Parkway NW, Suite 300
Boca Raton, FL 33487-2742

CRC Press is an imprint of the Taylor & Francis Group, an informa business

Visit the Taylor & Francis Web site at
http://www.taylorandfrancis.com

and the CRC Press Web site at
http://www.crcpress.com

Library of Congress Catalog Card Number: 79-4861

ISBN 13: 978-0-367-02125-2 (hbk)
ISBN 13: 978-0-367-17112-4 (pbk)

About the Book

Current and projected worldwide shortages of energy,
fertilizer, and irrigation water, coupled with a rapidly
expanding population, are prompting agricultural scientists
to seek, identify, and develop new crop species. Such crops
should be energy- and water-efficient, well adapted to marginal lands, suited to intensive culture, and productive of
marketable food, fiber, and other useful products. This book
introduces a variety of new crop species that display exceptional promise in these areas. While almost all of them are
in a precommercial stage of development, they are the focus
of active research in the United States and other countries.

Gary A. Ritchie is research project leader in the
Western Forestry Research Center of the Weyerhaeuser Company.

About the Series

The *AAAS Selected Symposia Series* was begun in 1977 to
provide a means for more permanently recording and more
widely disseminating some of the valuable material which is
discussed at the AAAS Annual National Meetings. The volumes
in this *Series* are based on symposia held at the Meetings
which address topics of current and continuing significance,
both within and among the sciences, and in the areas in which
science and technology impact on public policy. The *Series*
format is designed to provide for rapid dissemination of
information, so the papers are not typeset but are reproduced
directly from the camera-copy submitted by the authors, without copy editing. The papers are organized and edited by
the symposium arrangers who then become the editors of the
various volumes. Most papers published in this *Series* are
original contributions which have not been previously published, although in some cases additional papers from other
sources have been added by an editor to provide a more comprehensive view of a particular topic. Symposia may be reports of new research or reviews of established work, particularly work of an interdisciplinary nature, since the AAAS
Annual Meetings typically embrace the full range of the
sciences and their societal implications.

WILLIAM D. CAREY
*Executive Officer
American Association for
the Advancement of Science*

Contents

List of Figures

List of Tables

Preface

This volume had its genesis in Tacoma, Washington on a brisk February day in 1976. On that morning, a memo, which was to have a major impact on my life, found its way to my IN basket. Addressed "Dear Friend of AAAS," the memo called attention to the Association's Congressional Science Fellowship Program and invited member scientists to apply for a one-year Fellowship to work as special assistants in the United States Congress, while receiving a stipend from the AAAS.

Somewhat discouraged with my job at that time and needing a change of scenery, I elected to apply for one of the four Fellowships offered. Following much correspondence, preparation of an essay on a political issue, solicitation of three letters of reference, formulation of a pretend briefing paper to a fictitious Member of Congress, and a formal interview in Washington, D.C., I was informed by Dick Scribner of the AAAS that I had been selected as a Congressional Science Fellow for 1976–1977.

Approximately one year later, during my Fellowship tenure as a legislative assistant to Senator Pete V. Domenici of New Mexico, another important memo crossed my desk. It read: "TO: Gary, FROM: Charles, SUBJECT: Guayule."

Charles Gentry, himself a former Whitehouse Fellow, was the Senator's Chief Legislative Assistant at that time. A native New Mexican, Charles was alert to any potential actions Senator Domenici might take to attract economic development and benefits to his state. His memo invited my opinion as to whether or not guayule might offer such an opportunity.

Attached to the memo was a draft version of a National Research Council (NRC) report titled Guayule: An Alternative

Source of Natural Rubber. Being a resident of the rain
soaked Pacific Northwest, I was completely unfamiliar with
guayule -- a shrub of the parched Southwestern deserts.
Nevertheless, after having read the report I was firmly
convinced that a guayule rubber industry was just what New
Mexico needed and that Pete Domenici was the man who could
deliver it. Not unselfishly, I also saw guayule as my
ticket to gaining the Senator's personal attention -- a
reward jealously coveted by all Senate staff workers.

As I had hoped, Senator Domenici showed a keen and
immediate interest in the prospects for a guayule industry
in New Mexico and recognized an opportunity for a legislative
initiative which would result in the contribution of some
Federal dollars to the effort. He assigned to me the task
of drafting such a bill, which I did with great enthusiasm.
I called it the "Native Latex Commercialization Act of
1977," and Senator Domenici introduced it on Tuesday, March
29 -- the same day that the NRC released its final version
of the guayule report. In his introductory statement on the
Senate floor, Senator Domenici remarked:

> It is my firm belief, after considerable
> study of this matter, that a soundly
> conceived and aggressively conducted
> research and development program could
> result in a commercially viable Guayule
> latex industry in this country within 5
> to 10 years. This industry would not
> only free our Nation of its dependence
> on foreign sources of natural rubber,
> but it would also provide exceptional
> opportunities for agriculture, industry,
> and economic growth for depressed peoples
> -- especially American Indians -- now
> living in the arid and semiarid regions
> of the United States.

The bill was referred to the Committee on Agriculture
and Forestry where it died instantly.

Meanwhile, the AAAS annual meeting committee was solic-
iting ideas from members on suitable symposium topics for
its February, 1978 meeting in Washington. We Congressional
Science Fellows were asked to submit suggestions. My new
found interest in guayule, jojoba and other new crops along
with my association with Noel Vietmeyer, director of the
NRC's guayule study, prompted me to submit "New Agricultural
Crops" as a proposed topic. The committee liked it and I
began putting together a symposium. Noel's assistance was

invaluable for he had done extensive work in the area of new crop identification and development and, unlike myself, was personally acquainted with many active workers in the field.

The symposium was held on the morning of February 16, 1978. Four papers were presented: Felger on ancient Indian crops (see Chapter 1), Newell and Hymowitz on the winged bean (see Chapter 2), Hogan on jojoba (see Chapter 9) and Vietmeyer on guayule (see Chapter 8). It was extremely well received and we were asked by the AAAS whether we would be willing to have it published as a symposium volume. We agreed.

Subsequently, I was contacted by the AAAS, who requested that the scope of the volume be expanded by including several additional papers on crops which had not been addressed at the annual meeting. Again, I had to call upon my trusty friend Noel for a list of prospective crops and authors, which he graciously supplied. The new authors were contacted, agreements made, deadlines scheduled and re-scheduled, and this book was produced. My heartfelt thanks go out to all who have contributed.

Meanwhile, undaunted by the untimely demise of the Native Latex Commercialization Act, Pete Domenici outlined a new legislative strategy for guayule. It was clear to us that a second bill had no chance in the Agriculture and Forestry Committee, which was then weighted down with the burden of a six billion dollar farm bill. More importantly, Domenici did not serve on that committee and, therefore, had little if any influence over its legislative priorities.

Domenici was, however, on the Environment and Public Works Committee, where he was respected as an active and astute (albeit relatively new) member. In addition, the ranking member on the committee was Pete's very close friend and colleague, Howard Baker of Tennessee.

The new strategy: change the name of the bill to the "Guayule Rubber Demonstration and Economic Development Act of 1977" and reintroduce it. The Senate clerk would then automatically send it to the Environment and Public Works Committee, which has jurisdiction over legislation dealing with economic development. At least in Public Works, it would receive a fair hearing.

So the very same bill, but with a different name and some minor language adjustments, was reintroduced in the Senate on June 30, 1977 and was referred to the Committee on Environment and Public Works.

Shortly thereafter, my Congressional Science Fellowship terminated and I returned to the Pacific Northwest forests to resume my career as a plant physiologist. The route of the new guayule bill during the past year, I am told by my former colleagues, has been anything but smooth. While most of the details are obscure to me at this time, I do know that a companion bill was introduced in the House of Representatives by Congressman George Brown of California. Brown's bill was apparently identical to the Domenici bill but bore the original title of "Native Latex Commercialization Act." I am also told that the Department of Agriculture, while not particularly concerned with the original guayule bill, suddenly became highly interested in the new bill which authorized thirty million dollars to the Regional Economic Development Commissions rather than to the Agricultural Research Service. Although Domenici reprimanded the USDA representatives at the Senate hearings for their lack of interest in his original bill, a compromise was apparently struck in the House-Senate Conference Committee and the USDA was authorized half of the thirty million dollar grant.

The Conference Committee passed the bill and, as is custom, it then was given the title of the House version (which was the same as the title of the original Senate bill). It was passed without opposition by both Houses during the waning days of the 95th Congress.

On Saturday, November 4, President Carter signed into law the "Native Latex Commercialization Act of 1978."

Gary A. Ritchie

Olympia, Washington
December 18, 1978

About the Editor and Authors

Gary A. Ritchie, *a former AAAS Congressional Science Fellow, is currently a research project leader at the Weyerhaeuser Company Western Forestry Research Center, Centralia, Washington. A plant physiologist, he has authored many technical reports and a recent major review article on the physiological ecology of coniferous trees.*

W. P. Bemis, *professor in the Department of Plant Sciences at the University of Arizona, is a specialist in the field of plant genetics and evolution. He has recently published a number of papers concerning the genus* Cucurbita.

James W. Berry, *professor in the Department of Nutrition and Food Science at the University of Arizona, is a biochemist. His recent publications include papers on xerophytic cucurbits, sterculic acid, vinylated polysaccharides and glucose derivatives.*

James L. Brewbaker, *professor of horticulture and genetics at the University of Hawaii, has conducted tropical crop research in Hawaii, the Philippines, Thailand, India, and Colombia. He is president of the Hawaiian Academy of Sciences, former president of the Hawaiian Botanical Society, executive secretary of the Hawaiian Crop Improvement Association, and a fellow of the American Society of Agronomy. He is the author of one book and more than 120 papers on agricultural and biochemical generics, plant breeding, botany, and radiation biology.*

Laurie B. Feine, *a researcher at the Rodale Organic Gardening and Farming Research Center in Pennsylvania, specializes in plant ecology and his background is in environmental, population, and organismic biology.*

Richard S. Felger, *senior research scientist at the Arizona-Sonora Desert Museum and adjunct associate professor in the Department of Ecology and Evolutionary Biology at the University of Arizona in Tucson, conducts research on eth- nobotany and new agricultural systems in arid lands. His projects have ranged from studies of the ethnobotany of Seri Indians, to plant resource species of the Rio San Miguel Valley in Sonora and ethnobiology of sea turtles in the Gulf of Calfironia. He has published some 40 papers in these fields.*

Peter Felker, *assistant research plant physiologist in the Department of Soils and Environmental Sciences at the University of California-Riverside, is conducting research on leguminous trees as food and fuel sources in semi-arid lands and is the author of several related publications. He is currently principal investigator under a Department of Energy grant to evaluate mesquite for growth rate under controlled moisture stress.*

J. R. Goodin, *associate professor of biology at Texas Tech University, specializes in the physiology and biochemis- try of water and salt stress in plants. A participant in the Smithsonian Institution project on forage potential of arid zone halophytes, he has published one book and some 60 papers on salt tolerance in and utilization of arid land shrubs.*

Richard R. Harwood, *director of research at the Rodale Organic Farming and Research Center in Pennsylvania, specializes in multiple cropping and farming systems. He organized the Thailand National Sorghum Program (1967-72) and the International Rice Research Institute's Cropping Systems Program (1972-76) and was a member of the Vegetable Systems delegation to the People's Republic of China (1977).*

LeMoyne Hogan, *professor of horticulture at the Univer- sity of Arizona, has conducted research on new plant intro- duction and development and physiology of arid land plants. His work has involved vegetative propagation of jojoba, in- cluding tissue culture, and selection of jojoba for increased yield and cold hardiness.*

E. Mark Hutton *is currently visiting scientist in charge of pasture plant breeding with the Programma Ganado de Carne at the Centro Internacional de Agricultura Tropical in Colombia. He has many years experience breeding tropical pasture legumes, with particular emphasis on tolerance to acid soils. He was chief of the CSIRO Division on Tropical Crops and Pastures, where he bred the tropical legume Siratro and the new Cunningham Leucaena. He was president of the XI*

International Grassland Congress, and his awards include the 1968 Farrer Memorial Medal for distinguished contribution to Australian Agricultural Science. He has published some 100 papers in his area of expertise.

Theodore Hymowitz, *professor of agronomy at the University of Illinois at Urbana-Champaign, is a plant geneticist, with particular interest in germplasm resources and domestication of legumes. He has conducted field work in South America, Asia, and Australia, and he is a member of the NAS-NRC Ad Hoc Panel on Winged Bean and of the Winged Bean Steering Committee, and the editor-in-chief of the* **Winged Bean Flyer.**

C. S. Kauffman, *senior researcher with the Rodale Organic Gardening and Farming Research Center in Pennsylvania, is a specialist in horticulture and plant breeding and has published articles on the mode of inheritance of dwarfism in cucumbers. Formerly a manager of Tea Horticulture for Thomas J. Lipton, Inc., he has also been concerned with the economic feasibility of growing tea in the southeastern United States.*

C. A. Newell, *research associate in the Department of Agronomy at the University of Illinois at Urbana-Champaign, is a specialist in biosystematics and crop evolution.*

Donald L. Plucknett, *professor of agronomy at the University of Hawaii, has extensive experience in tropical agricultural research, in particular, agricultural diversification of marginal lands and neglected tropical crops. A former chief of AID's Soil and Water Management Division, he serves on several National Academy of Sciences panels and is president of the International Society for Tropical Root Crops. He was chairman of the NAS Vegetable Farming Systems Delegation to the People's Republic of China in 1977, and has published numerous articles and six books.*

Joseph P. Senft, *research scientist at the Rodale Organic Gardening and Farming Research Center in Pennsylvania, conducts research in membrane biophysics, active transport mechanisms, and amaranth nutrition. He has published several papers in his field of expertise.*

Noel D. Vietmeyer, *professional associate with the National Academy of Sciences, is a specialist in new crop development and science for developing countries. He was staff director for the NAS study of guayule and has directed NAS studies of winged bean, jojoba, leucaena, and other underexploited tropical plants.*

Charles W. Weber, *professor in the Department of Nutrition and Food Science at the University of Arizona, specializes in nutritional biochemistry. He has published papers on chemical analysis of various components of Buffalo gourd and nutrition evaluation of Buffalo gourd seeds.*

Introduction

Gary A. Ritchie

In a recent editorial in <u>Science</u> (1), S.H. Wittwer
speaks of the "next generation of agricultural research."
Identified in recent National Academy of Science-National
Research Council reports, this new wave of inquiry must move
us towards enabling plants and animals to more effectively
exploit present environmental resources in contrast to their
becoming increasingly dependent upon such scarce or non-
renewable resources as fossil energy, fossil water, chemicals
and capital. According to the thought-provoking editorial,
a new agricultural technology must embody (1) greater photosynthetic efficiency, (2) improved biological nitrogen
fixation, (3) new techniques for genetic improvement, (4)
more efficient nutrient and water uptake, and reduced losses
of nitrogen fertilizer from nitrification and denitrifica-
tion, and (5) more resistance to competing biological systems
and environmental stresses.

This symposium volume focuses on a technological strategy not specifically prescribed above but which holds immense
promise for world agriculture and the ultimate benefit of
man both in the near and long term. That strategy is the
domestication of wild plants and their introduction as new
agricultural crops.

While the plant species and varieties which are currently under serious consideration as potential crops may
number more than a hundred, this volume could necessarily
consider only a relative few. Nevertheless, these include
some of the most promising and exciting species covering a
broad spectrum of plant types, environmental requirements,
and products.

The volume begins with a review of the native plants
utilized by ancient peoples of the Southwestern Unitied
States. While most of the species described have only limited

promise as modern crops, others seem to hold great potential
and are the focus of current intensive development efforts
(e.g. mesquite, see Chapter 5).

Most of the species described in this volume are poten-
ial sources of human food. The winged bean (Psophocarpus
tetragonolobus), for example, has long been known by peoples
of the high rainfall tropics as a producer of protein and
oil. Many of its parts, including the leaves, flowers, pods,
stalks and roots, are edible and offer high food value. Its
promise lies in small family or community gardens as opposed
to large-scale agricultural operations.

Unlike the winged bean, amaranth (Amaranthus sp.) seems
well-adapted to large-scale, mechanized farming. This genus
is especially unique in two ways. Certain of its members,
the so-called vegetable amaranths, are best suited to the
production of leafy greens, while others are potential grain
producers. These are the only known non-monocotyledonous
grain species of any consequence.

Among the world's most problematical regions from an
agricultural standpoint are the arid and semiarid zones.
While these areas usually offer relatively fertile and
moderate textured soils, their universal dearth of water
(especially non-saline water) renders them unfit for cultivation of most conventional crops. Driven by the vast
potential of these regions and the increasing pressures upon
them, agricultural scientists from many nations are actively
seeking new crop species which are adapted to these harsh
conditions. Surprisingly, scientists are beginning to recognize that not only do many such species exist in the native
flora but also that several have extraordinary economic, as
well as agronomic, potential. The buffalo gourd (Cucurbita
foetidissima), for example, produces substantial yields of
vegetable oil, protein, starch and green biomass. It has
been utilized by native American peoples for centuries and,
with appropriate cultural methods, could become a major new
food and forage crop for semiarid regions.

The mesquite (Prosopis sp.) tree, while considered a
nuisance weed by many, is a marvelously productive and
useful plant for desert areas, thriving where few other
plants can subsist. Its pods contain 13% protein and up to
30% sucrose. Requiring little water and no tillage, mesquite
is a legume and, therefore, increases soil nitrogen. Its
potential for intensive culture as a human and animal food
producer has barely been tapped.

Another arid zone plant, *Atriplex*, aside from having a high forage potential, offers another unique feature. A halophyte, *Atriplex* is tolerant of saline soils and, therefore, may be cultivated in desert areas which are unfit for non-halophytic plants. Due to its ability to extract salts from the soil solution and concentrate them in the stem and leaves, plant scientists speculate that *Atriplex* may some day be used to "harvest" salts from the desert and thereby reclaim saline lands for agricultural production.

Not all the new crops discussed in this volume produce food and forage. The cassia (*Cassia* sp.) tree, for instance, has been known for perhaps 5,000 years in China as a source of spice (cinnamon) and essential oils. Recent research has suggested, however, that it may soon be possible to commercially produce *Cassia* outside of China on a large scale, thus creating new agricultural opportunities for many areas.

It is not widely recognized in the United States that natural rubber is an essential and strategic commodity, the entire supply of which comes from foreign producers. Practically all of the rubber used in large truck and aircraft tires in the United States is made from latex produced in Asian rubber plantations. Equally unknown is the fact that a small desert shrub called guayule *(Parthenium argentatum)*, native to Mexico and the southwestern United States, could be the source of a domestic natural rubber industry. This shrub can contain, up to 25% by weight, a latex comparable in every way to Asian latex. In fact, during World War II a significant quantity of guayule rubber was produced by the government in a massive effort to replace rubber supplies captured by Japanese forces in the South Pacific. There are many strong indications that the time may have come for a renewed large-scale interest in guayule rubber.

In addition, it appears that our national effort to save the sperm whale from extinction might be aided by a small, previously little-known desert shrub known as jojoba *(Simmondsia chinensis)*. Aside from its unique and highly prized oil, the sperm whale has only marginal commercial value. The oil, however, is extraordinarily useful, being a critical ingredient in many pharmaceuticals, lubricants, waxes and cosmetics. Due to a massive demand for the oil and the concomitant depletion of sperm whale populations by overhunting, the oil has become increasingly scarce and very expensive.

Jojoba beans contain high concentrations of a liquid wax which has properties very similar to those of sperm

whale oil. In fact, jojoba oil can be substituted for sperm
whale oil in practically every one of its many uses. The
potential for jojoba agriculture in Israel, Mexico, Australia
and the warmer areas of the American Southwest is now under
intensive investigation.

Finally, one of the fastest growing useful plants known
to man is the tropical tree, Leucaena (*Leucaena leucocephala*).
A leafy evergreen native to Mexico and closely resembling the
mimosa, leucaena plants can be grown for cattle forage, wood,
paper pulp and fertilizer.

As a forage crop, shrub-type leucaena varieties have
leaves high in protein and similar in quality to alfalfa. It
is particularly palatable to dairy cows, beef cattle and
water buffaloes. Cattle feeding on leucaena have, shown
weight gains comparable to those of cattle feeding on the
finest pastures anywhere.

Its greatest potential contribution, however, may be in
reforesting large areas of the tropics which have been
denuded through uncontrolled agriculture and logging. Able
to attain heights of 20 meters in from six to eight years,
leucaena could help to relieve critical shortages of fuelwood
in developing countries.

The plant kingdom contains somewhere on the order of
500,000 species. Of this vast germplasm resource only about
1,000 species, or 0.2% of the total, are exploited for man's
benefit. In the United States, about 30 species make up 95%
of our diet. Additionally, virtually every major food crop
used today has been known and cultivated for at least two,
and perhaps as long as ten, millennia (2).

While domestication of wild plants is, admittedly, a
costly and time consuming process, the tools, finances, and
human skills available are fully adequate to the task. To
what more worthy enterprise could these resources be com-
mitted?

References and Notes

1.   S.H. Wittwer. Science, 199, 375 (1978).

2.   C.L. Wilson, W.E. Loomis and H.T. Croasdale. Botany,
     3rd edition. Holt, Rinehart and Winston. N.Y. (1962).

1

Ancient Crops for the Twenty-first Century

Richard S. Felger

Abstract

All over the world there are myriad plant and animal species which can be adapted to new agricultural systems. These species, which once supported indigenous peoples, include many life-forms which can be developed for modern crops specifically adapted to local conditions. Examples of potential new crops taken from the repertoire of major food resources of various Indian groups in the Sonoran Desert alone include: desert trees and shrubs, desert palms, columnar cacti, desert ephemerals, saltgrasses, and seagrasses. The potential for new crops has hardly been tapped.

Introduction

Worldwide there probably are thousands of plants which can be developed for major "new" crops. The term "new crop" is used here to indicate one new to modern agriculture. What may be a new crop for us includes both wild species and cultivated domesticates which have been utilized by native peoples since ancient times. These include life-forms (=growth-forms) and species of plants which seldom have been considered as agriculturally important.

By utilizing diverse indigenous crops from the different parts of the world, we can develop agriculture specifically adapted to local environmental conditions. Crops can be designed to fit the environment, rather than the environment being modified to suit the crop (1). Conservation of energy and resources would be substantial. Examples are presented from the Sonoran Desert region of southwestern North America. The present discussion centers on potential food crops derived from seed plants (spermatophytes), although similar methodologies can be

applied to non-food crops as well as animals and lower
plants.

Diversity

The most important food crops of the world number only
several dozen species, and seven of these keep the majority
of humanity from starvation: wheat, rice, maize (corn),
barley, soybean, common bean, and potato. Most of these are
derived from temperate or tropical ancestors and have
relatively large water requirements. In contrast, it is
estimated that on the order of 3,000 species, representing
approximately 1.5% of the world flora of seed plants, may
have potential as major food crops for the world (2). These
include species adapted to most of the major biomes of the
world. This multitude of species could be developed to
utilize many different niches such as previously irrigated
desert farmland, saline soils and even sea water. New crop
agriculture could increase niche diversity and life-form
diversity as well as species richness in man-made
environments.

Even if a fraction of the estimated thousands of major
new food crops are ultimately successful, the world
repertoire of major food crops will be increased by orders
of magnitude and there could be a marked improvement in
world nutrition and biological diversity. The impact on
global events could be profound.

Diversity provides redundancy in a biological
community. If one system fails, the other is available as a
backup. Agricultural diversity can also be redundant in
many aspects including protein, oil, and carbohydrate
production, as well as utilizing plural inputs such as
different niches, geographic locations, economies, climates,
and human cultures.

In recent times there has been drastic loss of
agricultural diversity (3,4). This agricultural
specialization has certain agronomic benefits as long as
energy and the resource base remain to sustain the system.
But because of the need to conserve energy and biological
resources there is an adaptive value in expanding genetic
diversity. Thus, under present and foreseeable future
conditions, overall increases in crop diversity will lead to
more efficient agricultural schemes.

Methodology

Initial research on new crops should include systematic

surveys of entire regions for potentially significant crops. A broad-spectrum survey of this nature would include analysis and interpretation of ecological, ethnobotanical, and nutritional information. Of primary significance is an understanding of ecological parameters and limiting factors, such as temperature and drought, and how they affect yields of edible or usable products. Considerable insight into potential yields under cultivated conditions can be gained by studying yields under natural conditions (or crop ecology in the case of native domesticates). If factors of this nature indicate a particular plant or group of plants has promise, then further research and development should be carried out under cultivated conditions. Such R & D would include studies on genetic improvement, horticulture, crop yield, potential pathogens and pests, harvesting, storage, shipping, marketing, and other agronomic considerations. Nutritional characteristics of various new crops are discussed in other chapters of this volume.

Until now, all major crops have been domesticated by ancient or "primitive" peoples. The buffalo gourd, currently being domesticated by agricultural scientists, may become one of the first major new food crops of our times (Bemis, this volume). In the past, efforts to develop new crops generally have been scattered, with agricultural scientists often working in relative isolation and specializing on a single taxon.

It may be that a particular candidate plant ultimately is not utilized, but as the pool of potential new crops is increased the chance of finding a winner likewise increases. Furthermore, information gained from one candidate may apply to other, related potential crops. For example, although buffalo gourd (<u>Cucurbita</u> <u>foetidissima</u>) proved to be the better crop, much pertinent information was derived from extensive study of the coyote gourd (<u>C</u>. <u>digitata</u>) (Bemis, this volume).

R & D on the buffalo gourd (Bemis, this volume) and jojoba (Hogan, this volume) clearly demonstrates the value of multiple products from a single crop. This is particularly important during the initial economic stages of a new crop. For example, jojoba has considerable value as a liquid wax replacement for sperm whale oil. However, it is not yet economically viable, based on market price for the liquid wax. But the cosmetic industry is able to afford the high price necessitated by initial scarcity of the product. Another consideration is the problem of introducing a new food into a culture. Although it is often argued that people do not readily accept new foods, there are well documented cases for widespread acceptance of major new foods. Witness the soybean in our culture, and the agricultural aftermath of the Columbian exchange (5). Nevertheless there are legal and cultural considerations involved in introducing new foods. In many cases it would probably be best to initially utilize a new food crop as animal feed while solving the problems of general acceptance for human consumption.

There is an urgent need to place greater priority on recording ethnobotanical information from non-literate peoples. The importance of ethnobotanical investigations of large-scale ecosystems, such as the Sonoran Desert, cannot be overestimated. Native knowledge of indigenous wild and cultivated plants is valuable for the design of new crops. Much of this knowledge has already vanished unrecorded; that which remains is being lost at an accelerated rate.

Ethnocentric agricultural missionary effort from technologically advanced countries continues to dominate and stultify the use of alternative, and perhaps locally superior, plant materials and systems, especially in non-temperate regions. When acculturation occurs, as it invariably does, native knowledge usually dies with the elders of the community. Anthropologists lacking a strong background in the botany or economic botany of the region, and botanists lacking anthropological or linguistic sophistication are often unable to record, or recognize significant ethnobotanical information. Ethnobotany can provide access to thousands of years of knowledge, including subtle information that might otherwise be lost (6).

Sonoran Desert Crops

Since approximately one-third of the land surface of the world is arid or semi-arid (7), it behooves us to look for crops specifically adapted to dry lands. Seeds and other storage organs of arid land food plants commonly contain high concentrations of food energy (6, 8, 9), while often being drought resistant or able to produce crops in relatively short growing seasons.

In the Sonoran Desert as well as elsewhere, major carbohydrate foods are derived from certain seeds, fleshy non-seed fruit parts such as mesocarp tissue, and various storage organs such as thickened stems and roots or tubers. Major sugar sources include non-seed portions of fruits which have evolved as reward-stimuli for animal dispersal of the seeds. These sugary foods do not require cooking or processing. In contrast, many carbohydrate-rich storage organs often contain starch and heat-labile toxins, and therefore need to be thoroughly cooked (10, 11, 12, 13).

The important sources of vegetable protein and oil are seeds. Although seeds of arid land plants tend to be small, they are notably high in protein and/or oil content (6). However, just because protein content is high does not necessarily mean that the amino acid spectrum is nutritionally favorable. Almost all of the major native plant-derived foods of the Sonoran Desert, particularly seeds, seem to be relatively free of toxic substances, at least when cooked. Many seeds tend to be undigestible unless cracked or broken, and native desert peoples usually parch the seeds to facilitate grinding them into flour. By greatly increasing the surface area of food particles, less time and water is needed for cooking (12, 13, 14).

The relatively low-energy vegetative parts (leaves, stems, and roots) of many desert plants are high in secondary compounds, such as alkaloids and terpenes. This is seen as an adaptation enhancing predation avoidance. However, people have long used these secondary compounds for an array of utilitarian and medicinal purposes. Thus it is not surprising to find that the major plant-derived foods of a desert people such as the Seri Indians of Sonora, Mexico, consist largely of seeds and fruits while the majority of the more than 100 species of plants used for medicinal purposes are prepared from vegetative organs (14, 15).

The Sonoran Desert embraces approximately 310,000 sq km and supports a flora of about 2500 species of seed plants (16). Felger and Nabhan (2) report "...that about 18% of this flora or about 450 species (about 375 native species and about 75 naturalized species) were utilized for food by the various native peoples in this region. More than 10% of the 375 edible species have been utilized as major food resources." A sampling of important Sonoran Desert food plants which seem to have agronomic promise are listed below. Following this list is a brief discussion of selected potential food crops in three major families of Sonoran Desert plants.

Desert Ephemerals
  cool weather species:
    Descurrainia pinnata, tansy mustard: seed
    Lepidium lasiocarpum, peppergrass: seed
    Lesquerella spp., bladder-pod: seed
    Plantago insularis, Indian wheat: seed
  warm weather species:
    Amaranthus spp., careless weed: seed and leaves

Mentzelia spp., blazing star: seed
Phaseolus acutifolius, tepary: seed
Proboscidea parviflora, devil's claw: seed
Panicum sonorum, saui: grain
  non-seasonal species:
  Oligomeris linifolia, linear-leaved cambess: seed
Root Perennials
Amoreuxia palmatifida, saiya: root, leaves, fruit,
  and seed
Cucurbita digitata, coyote gourd: root, seed
Cucurbita foetidissima, buffalo gourd: root, seed
Jarilla chocola, chócola: root, fruit
Phaseolus ritensis, cocolmeca: seed
Salpianthus macrodontus, guayavilla: root
Trees and Shrubs
Acacia cochliacantha, boat-spine acacia: seed
Cercidium floridum, blue palo verde: seed
C. microphyllum, foothill palo verde: seed
Lysiloma divaricata, mauto: seed
L. watsonii, tepeguaje: seed
Olneya tesota, ironwood: seed
Pinus cembroides, Mexican pinyon: seed
Pithecellobium dulce, guaymuchil: fruit (mesocarp)
Prosopis glandulosa, mesquite: fruit (mesocarp) and
  seed
P. velutina, mesquite: fruit (mesocarp) and seed
Quercus emoryi, Emory oak: seed
Palms
Brahea armata (=Erythea), Blue hesper palm: seed
B. edulis (=Erythea), Guadalupe fan palm: fruit
  (mesocarp) and seed
Washingtonia filifera, desert fan palm: fruit
  (mesocarp) and seed
W. robusta, desert fan palm: fruit (mesocarp) and seed
Cacti
  columnar cacti:
  Lemaireocereus thurberi, organ pipe cactus: fruit
    (pulp and rind)
  Machaerocereus gummosus, pitahaya agria: fruit
    (pulp)
  Pachycereus pecten-aboriginum, echo: fruit (pulp) and
    seed
  P. pringlei, cardón: fruit (pulp) and seed
  prickly pears and chollas:
  Opuntia fulgida, jumping cholla: fruit (excluding
    seed)
  O. phaeacantha, prickly pear: fruit (excluding seed)
  small cacti:
  Mammillaria microcarpa and M. spp., pincushion
    cactus: fruit including seed

Saltgrasses
    Distichlis palmeri, Palmer saltgrass: grain
Seagrasses
    Phylospadix torreyi, surfgrass: seed
    Zostera marina, eelgrass: seed

Cactaceae - Cactus Family

There are about 2,000 species of cacti native to the
New World (17). About 145 species occur in the Sonoran
Desert (16), and most of these were used for food by the
local peoples. The family includes a great variety of
growth-forms, ranging from thimble-sized to towering giants
Because of a high degree of morphological plasticity and
hybridization potential within the family (6, 18), different
cacti can be chosen and developed to fit a wide range of
environmental conditions within warm arid and semi-arid
lands. However, beyond the arid tropical regions, such as
the northern part of the Sonoran Desert, freezing
temperatures in winter can be expected to be a major
limiting factor. As with most tree or long-lived perennial
crops, generation time can be long, ranging from about half
a dozen years to decades for the slower-growing giants.

A limited number of cacti have dry, inedible fruit.
All others have fleshy fruit which can be eaten fresh,
although there is considerable variation in taste. The
fleshy fruit usually contains many small seeds high in
protein and oil content. However, the seeds of opuntia and
related genera are not usually edible. Three major growth-
forms or size-classes of Sonoran Desert cacti may be of
agronomic interest: the dwarf pincushion cacti, the shrub-
sized prickly pears and chollas, and the large columnar
cacti.

There are several hundred species of Mammillaria, a
small and usually globose-shaped cactus (17). Within the
Sonoran Desert there are about 40 species (16). All have
fleshy and spineless fruits which are commonly 2 to 2.5 cm
long. The fruit is sweet and often tart, and the seed high
in protein and oil content. The fruit is relished by all
who sample it. With the enormous hybridization potential,
considerable genetic improvement should be possible. It is
recommended as a strawberry-like specialty crop. The
succulent fruit should ship well. Quantitative yields are
unknown, but each stem, about 10 to 20 cm in diameter, can
produce several dozen fruit at least once a year.

The genus Opuntia, as it is usually interpreted,
includes the prickly pears (subgenus Platyopuntia) with

flattened leaf-like stems, and the chollas (subgenus
Cylindropuntia) with cylindrical stems more or less round
in cross-section. The stems of most opuntias do not contain.
toxic substances, and some are edible if still young and
tender.

The different species of prickly pear exhibit great
variation in size and ecological distribution. Interesting
hybrids could probably be made between certain Sonoran
Desert prickly pears such as Opuntia phaeacantha with
sweet, juicy fruits, and the large domesticated Mexican
prickly pear, O. ficus-indica. The latter is known to
hybridize readily with wild prickly pears in the Southwest
(19).

Perhaps one of the most bizarre potential new crops is
the jumping cholla, Opuntia fulgida. The Seri Indians
gather the succulent, greenish-colored fruit and consume it
as a vegetable, either fresh or cooked. The fruit is 2.5 to
4 cm long, roundish, spineless but with glochids, often
seedless, and produced in pendulent chains from the major
branches. The flavor is pleasantly tart and slightly sweet.
The fruit should hold up very well under short-term storage
and shipping. The populations and individual plants show
considerable variation. Some have branches densely covered
with spines and others have shorter and fewer spines. The
spines are obnoxious and if developed as a commercial crop,
it would be desirable to select for a spineless condition.

The following discussion on columnar cacti is adapted
from an earlier report by Felger and Nabhan (2). These
cacti have one or more stems which may reach shrub to tree
heights. The larger ones may weigh many tons. Among the
several hundred species in the Americas, there is a wide
range of ecological tolerances. Some are extremely drought
resistant, such as Pachycereus pringlei occurring with an
average annual rainfall of less than 200 mm.

Sonoran Desert columnar cacti which seem to have
promise as dry-land crops are: Lemaireocereus thurberi
(organ pipe), Machaerocereus gummosus (pitahaya agria),
and Pachycereus pringlei (cardón) and P. pecten-
aboriginum (echo). These cacti provided important food
resources and wines for Sonoran Desert peoples, and together
with Carnegiea gigantea (saguaro), are still harvested for
their fruits (20, 21).

Fresh weight of the fruit is about 150 g or more, and
50 to 100% of the fruit is edible. There is usually a
sweet, sugar-rich juicy pulp containing many small protein

and oil-rich seeds. For example, seeds of card6n contain
22.6% protein and 32.2% edible oil (14). Sugar content and
quality are exceedingly high (22, 23). Each species has a
distinct flavor, and these fruits should be suitable for
wines as well as non-alcoholic beverages for hot, arid
tropical regions.

Some columnar cacti have spineless fruit, but others
are spiny although the spines usually fall off when the
fruit is mature. The fruit can be consumed fresh, or it can
be dried or preserved. Organ pipe is esteemed for its very
sweet, succulent fruit. Unlike other Sonoran Desert
columnar cacti, the entire fruit is edible. Pitahaya agria
has a delicious sweet but very tart flavor. The seeds of
both organ pipe and pitahaya agria are small and were not
processed for food. Card6n has both a desirable sweet pulp
and seeds large enough to warrant processing. The seeds are
parched and ground into an oily paste which tastes somewhat
like peanut butter when salted, but is grey instead of
brown.

Fabaceae - Legume Family

Many of the most important economic plants of the world
are legumes, e.g., acacia, alfalfa, beans, peanut, peas, and
soybean. Because many are nitrogen-fixing, they are
especially significant agronomically and in natural
ecosystems. Legumes are particularly abundant in warm arid
or semi-arid regions such as the Sonoran Desert. Because of
their presumed nitrogen-fixing qualities, the major Sonoran
Desert legumes appear to be the primary source of nitrogen,
and hence amino acids and protein for the desert. The
common, large desert legumes provide major food inputs for
animals, and in earlier times were harvested by the
Amerinds. There are about 282 species of legumes known from
the Sonoran Desert (16) and about 40 of these were utilized
for food by the native peoples.

Largely due to the work of Gary Nabhan of the
University of Arizona, there is renewed agronomic interest
in the tepary (Phaseolus acutifolius). This ephemeral or
annual bean includes both wild and domesticated varieties
indigenous to southwestern North America which have been
utilized as food by native Americans for more than five
millenia. Seed production for domesticated teparies can be
high, ranging up to 2,000 to 4,630 kg/ha depending on soil
moisture. The seed stores well, has a high protein content
(21.1% to 32.5%), and is nutritionally comparable or
superior to most economic legumes (8, 24).

Table 1. Yields of selected desert legume trees
sampled from natural populations in southern Arizona.

species	mean weight of seed (or pod) (g)	maximum yield tree/year (kg) [and tree diameter (m)]	mean yield tree/year (kg) [and mean tree diameter (m)]	n of trees
Cercidium microphyllum	0.15	8.9 (7.5)	1.4 (4.1)	29
Cercidium floridum	0.18	8.1 (7.5)	1.2 (4.2)	22
Olneya tesota	0.15	2.7 (9.1)	1.7 (5.1)	9
Prosopis velutina	2.26 (pod)	9.1 (7.0)	4.1 (6.2)	20

Table 2. Yields of selected desert legume trees sampled
from natural populations in southern Arizona projected to
orchard configurations.

species	trees/ha	optimal diameter of trees (m)	seed (or pod) weight/tree (kg)	kg/ha/yr
Cercidium microphyllum	336	3.64	1.20	401.5
Cercidium floridum	336	3.64	0.88	295.8
Olneya tesota	165	6.5	1.7	280.5
Prosopis velutina	100	9.0	4.1 (pod)	410.0

The seeds and pods of certain Sonoran Desert legumes were among the most important food resources for the Indian peoples of the region. Among the genera which include species of primary significance are Acacia, Cercidium (palo verde), Lysiloma, Olneya (desert ironwood), Pithecellobium, and Prosopis (mesquite and screwbean). These plants were also utilized for a wide variety of non-food resources. Mesquite produces pods with abundant carbohydrate-rich mesocarp. The seeds were also sometimes used, but are relatively small and difficult to process. Screwbean (Prosopis pubescens) and guaymuchil (Pithecellobium dulce) pods are harvested primarily for the sweet mesocarp, but the relatively small seeds are not used. Most of the remaining species yield pods with relatively large seeds with no mesocarp.

Among the species which appear to be most interesting as potential new crops and those for which the most data is available are Cercidium floridum (blue palo verde), C. microphyllum (foothill palo verde), Olneya tesota (ironwood), and Prosopis glandulosa and velutina (mesquite). Studies of seed and pod yields of these species under natural conditions are being conducted at the Arizona-Sonora Desert Museum by Dennis Cornejo and Felger. Several hundred trees in the Sonoran Desert have been sampled during the past three years. Some preliminary results of these studies are shown in Tables 1 and 2. Extrapolations to simulated orchard conditions indicate potential yields of 280 to 410 kg/seed or pods/ha. These estimates include individuals with zero yields in the year sampled. In contrast some trees, notably certain populations of mesquite, have high yields on successive years, presumably because of the partial phreatophytic habit (25). However, for many desert legume trees, a year of high production is often followed by one of reduced yield. By monitoring the same trees each year, factors which influence production are being determined. With selection, genetic improvement, and agricultural conditions it should be possible to increase yields at least several times, which would be competitive with modern crops.

The North American species of Prosopis in the section Algarobia, numbering about half a dozen species, are collectively known as mesquite. In addition there are about two dozen species in South America (26). "From early pre-historic times until recent years, mesquite has served native peoples in southwestern North America as a primary resource for food, fuel, shelter, weapons, tools, fiber, medicine, and many other practical and aesthetic purposes. Every part of the plant is used. Utilization of mesquite

was the common denominator among the diverse peoples of the
arid southwestern lowlands...Because mesquite is such an
important and unfailing resource, it came to figure in the
everyday life of these peoples from cradle to grave" (27).

Mesquite possesses numerous features which should be
advantageous for developing it as a new crop. As with palo
verde and ironwood, the pods ripen nearly simultaneously,
facilitating the harvest. The pods are indehiscent, large,
and fall when fully ripe. The pods were dried, or sometimes
parched, and the mesocarp extracted by pounding them in a
mortar. The resulting flour was consumed either as a gruel
(atole), or mixed with water, formed into a cake and dried.
Significantly, no cooking is necessary, which should be very
important in many fuel-starved parts of the world.
Approximately 50% of the pod by weight consists of mesocarp
tissue and 15% consists of seed (28).

There is considerable variation in the sweetness of the
pods. For example, in Baja California the pods are usually
bitter in comparison with those from Sonora (29).
Furthermore, the Seri Indians know of certain groves with
high yielding and superior tasting fruit (30).

Although the seeds are relatively small, they are high
in protein content (Felker, this volume). A major drawback
is that the seeds are encased in a stony endocarp or pit
which is difficult to crack. However, earlier peoples
devised a stone tool, called a gyratory crusher, to process
the pods and crack the endocarp to free the seed (1, 27).
As with many legumes, species of Bruchus, a small weevil-
like beetle, are a potentially serious pest (31). Prompt
harvesting and processing may alleviate the problem. The
Seri Indians heat-treated the pods, primarily to facilitate
pounding and grinding, but this should be an effective means
of combating bruchid damage (30).

Among the approximately 10 species of Cercidium (17),
several of the larger-seeded species, in particular C.
floridum (blue palo verde) and C. microphyllum (foothill
palo verde), seem to have agronomic promise. The pods are
dry when mature, without mesocarp, and essentially
indehiscent. The seeds are relatively large and the yields
high (Tables 1 and 2). Protein content for C. floridum is
reported to vary from 27.5% to 54.4% (8, 9). Natural
hybrids and infraspecific variation are known (32, 33). The
seeds are very hard when dry and mature, and are parched and
ground into flour. The Papago Indians of southern Arizona
prefer foothill palo verde, saying that the seeds of blue
palo verde do not taste good. When immature and green, the

seeds are sweet and tender. Blue palo verde usually grows along dry, desert watercourses, and foothill palo verde is locally abundant on dry desert slopes.

Olneya tesota, locally known as ironwood, is a monotypic genus endemic to the Sonoran Desert. Seed size seems to increase southward, indicating geographic variation with some potential for genetic improvement. Yield is comparable with that of other major desert legumes (Tables 1 and 2). Ironwood is sensitive to frost, tolerating about as much freezing weather as does citrus. Flowering and fruiting usually do not occur every year, with severe frost certainly being an important limiting factor, and soil moisture probably being another factor (16). The pods are variously indehiscent or tardily dehiscent, and tend to fall when ripe.

The seeds were gathered by various Indian groups. The Seri Indians boil the seeds and discard the water, and cook them again in water to soften them. The water is changed to remove a bitter taste. The seed coat floats to the surface and is discarded. This is the only food resource which Sonoran Desert Indians cooked in a "second water". The cooked seeds are eaten whole or ground and salted. The Seri say the ground seeds taste like peanut butter (14). The bitter taste apparently derives from a free amino acid, canavanine, which also occurs in various other legumes such as the jack bean (34). Olneya seems to have significantly less bruchid infestation than do other major desert legumes. Potential drawbacks to ironwood as an agronomic crop are its relatively slow growth, sharp spines, and low potential for hybridization.

Poaceae - Grass Family

This family includes some of the world's most important crops: wheat, rice, maize (corn), barley, oats, sorghum, millet, sugar cane, etc. The flora of the Sonoran Desert includes about 167 species of grasses (16). Amerinds in the Sonoran Desert utilized about 30 to 35 of these species for subsistence.

One of the most interesting Sonoran Desert grain crops is Palmer saltgrass (Distichlis palmeri), known only from the delta region of the Colorado River. It was harvested in great quantity by the Cocopa Indians prior to the construction of upriver dams. As evidenced by recent linguistic information gathered by Amadeo Rea and Felger, the Cocopa may have had an improved variety which now may be extinct. The last harvest of this grain by the Cocopa was in the early 1950s. The grain is about the size of that of wheat. The plant is a dioecious perennial, and tolerant of drought and saline water (2, 6, 11, 35).

The hybridization potential should be high, with a dozen congeneric species and several other related genera, some of which are also tolerant of saline conditions (2). The promise of a salt and drought resistant grain crop make this a potentially valuable resource.

Conclusion

Few of the plants mentioned here as potential crops can, at this time, be competitive with modern crops. However, the development of new crops such as these needs to be pursued. Ethnobotanical and ecological studies of the prospective target species or taxon can shorten the development time. Study of various ecological parameters can provide critical information for the adaptation of a new crop, e.g., yields, limiting factors such as moisture and temperature sensitivity, reproductive strategies, or strategies for predation avoidance or resistance. If ecological and ethnobotanical studies indicate promise, it will be necessary to initiate long term research, since short term projects will probably be inconclusive (21). This research should apply modern techniques of crop development, and ideally would be integrated with a world agriculture strategy spanning the coming century.

This prospective is not without problems. The long range research programs required will be impracticable without substantial funding. Unfortunately, support for new crops research remains minimal. In the United States there is essentially no funding available to the general scientific community for research on new food crops. Nevertheless, the potential benefits of new crop agriculture are great. We can expect that this field is in its infancy.

Acknowledgments

A contribution from the Research Department of the Arizona-Sonora Desert Museum, Tucson, AZ 85704. Research supported in part by the National Science Foundation (BNS-77-08-582) and the Ray Foundation of Montana. I thank the following colleagues for valuable input and contributions: Martha Ames, Dennis O. Cornejo, Sylvia Earle, Donna Howell, Clayton J. May, Jean Mayer, Mary Beck Moser, Gary P. Nabhan, Amadeo Rea, Jacqueline Soule, Charles Stigers, and Barbara Tapper.

References

1. R.S. Felger and G.P. Nabhan, Ceres 9, 34 (1976).
2. R.S. Felger and G.P. Nabhan, in Social and Technological Management in Dry Lands, N. Gonzalez, Ed. (Westview Press, Boulder) pp. 129-149.
3. D.H. Janzen, Science 182, 1212 (1973).
4. National Academy of Sciences, Conservation of Germplasm Resources (Washington, D.C., 1978).
5. A.W. Crosby, Jr., The Columbian Exchange: Biological and Cultural Consequences of 1492 (Greenwood Press, Westport, 1972).
6. R.S. Felger, in Priorities in Child Nutrition in Developing Countries, J. Mayer and J.W. Dwyer, Eds. (UNICEF, New York, 1975), pp. 373-403.
7. W.G. McGinnies, B.J. Goldman, P. Paylore, Eds., Deserts of the World (University of Arizona Press, Tucson, 1968).
8. F.R. Earle and Q. Jones, Econ. Bot. 16, 221 (1962). Q. Jones and F.R. Earle, Econ. Bot. 20, 127 (1966).
10. E.F. Castetter and W.H. Bell, Pima and Papago Indian Agriculture (University of New Mexico Press, Albuquerque, 1942).
11. E.F. Castetter and W.H. Bell, Yuman Indian Agriculture (University of New Mexico Press, Albuquerque, 1951).
12. C. Niethammer, American Indian Food and Lore (Macmillan Publishing Co. Inc., New York, 1974).
13. R. Gasser, Hohokam Subsistence (Archeological Report No. 11, USDA Forest Service, Southwest Region, Albuquerque, 1976).
14. R.S. Felger and M.B. Moser, Ecol. Food Nutr. 5, 13 (1976).
15. R.S. Felger and M.B. Moser, Econ. Bot. 28, 414 (1974).
16. F. Shreve and I.L. Wiggins, Vegetation and Flora of the Sonoran Desert (Stanford University Press, Stanford, 1964).
17. J.C. Willis, A Dictionary of the Flowering Plants and Ferns, 8th edition (Cambridge University Press, Cambridge, 1973).
18. N.L. Britton and J.N. Rose, The Cactaceae (Dover Publications, Inc., New York, 1963).
19. L. Benson, The Native Cacti of California (Stanford University Press, Stanford, 1969).
20. E.F. Castetter and W.H. Bell, Ethnobiological Studies in the American Southwest. IV. The aboriginal utilization of the tall cacti in the American Southwest (Univ. N. Mex. Bull. 307, Biol. Ser. 5, 1937).
21. R.S. Felger and M.B. Moser, Kiva 39, 257 (1974).

22. R.A. Greene, J. Chem. Educ. 13, 309 (1936).
23. R. Pant, Curr. Sci. 42, 721 (1973).
24. G.P. Nabhan and R.S. Felger, Econ. Bot. 32, 2 (1978).
25. B.B. Simpson, Ed., Mesquite (Dowden, Hutchinson and Ross, Inc., Stroudsberg, Penn., 1977).
26. A. Burkart and B.B. Simpson, ibid., pp. 201-215.
27. R.S. Felger, ibid., pp. 150-176.
28. G.P. Walton, Nutritional Value of the Beans of Prosopis juliflora, Ceratonia siliqua and Gleiditschia triacanthos (U.S. Dep. Agric., Bull. 1194, 1923).
29. M. León-Portillo, Historia Natural y Crónica de la Antigua California (Universidad Nacional Autónoma de México, Mexico City, 1973).
30. R.S. Felger and M.B. Moser, Kiva 37, 53 (1971).
31. J.M. Kingsolver, C.D. Johnson, S.R. Swier, A.L. Teran, in Mesquite, B.B. Simpson, Ed., (Dowden, Hutchinson and Ross, Inc., 1977), pp. 108-122.
32. A.M. Carter, Proc. Calif. Acad. Sci. 4th Ser. 40, 17 (1974).
33. C.E. Jones, Evolution 32, 189 (1978).
34. G.A. Rosenthal, Q. Rev. Biol. 52, 155 (1977).
35. G. Vasey, Garden and Forest 2, 401 (1899).

2

The Winged Bean as
an Agricultural Crop

C.A. Newell and Theodore Hymowitz

Abstract

The genus Psophocarpus is composed of about 9 species,
8 of which are indigenous to Africa. The ninth species is
P. tetragonolobus or the winged bean. The plant is a
climbing perennial with large 4-angled pods and is grown on
a trellis as an annual. It is cultivated in the high rainfall tropics and subtropics of Southeast Asia, Oceania and
Africa. The seeds have approximately the same quantity and
quality of protein and oil as that of the soybean. In addition, the tubers have about 5 to 20% protein. In some
countries, leaves, flowers, and young pods form a minor part
of the human diet. There appears to be considerable morphological and chemical variation within the domesticate.
However, seed collections are few and a major effort should
be made especially in Papua New Guinea and Indonesia to
obtain additional germplasm. The greatest potential for the
winged bean is as a protein crop for those high rainfall
areas which traditionally depend on staple crops such as
yams and cassavas which are low in protein. Unfortunately,
relatively few agronomic, nutritional, and plant breeding
studies have been conducted on the winged bean.

Introduction

In the past 4 years, the winged bean - Psophocarpus
tetragonolobus (L.) DC. - a large seeded legume, has received
a great deal of attention from the scientific community (1,
2, 3). With peoples of many tropical and subtropical
developing countries facing a stark future regarding supplies
of food, the rediscovery of the winged bean comes at an
appropriate time. The winged bean is a multi-purpose edible
cultigen that grows easily, quickly and yields abundantly.
Its green pods, leaves and seeds are rich in protein and
vitamins. The tubers are unique in that they contain a high

See Note, page 37.

Table 1.   Species of the genus Psophocarpus and their
distribution.

Species	Distribution
1. P. grandiflorus Wilczek	Ethiopia, Uganda, Zaire
2. P. lancifolius Harms	Burundi, Kenya, Malawi, Nigeria, Rhodesia, Rwanda, Tanzania, Uganda, Zaire, Zambia
3. P. lecomtei Tisserant	Central African Republic, Zaire
4. P. lukafuensis (DeWild) Wilczek	Zaire, Zambia
5. P. monophyllus Harms	Guinea Bissau, Guinée & Côte d'Ivoire, Mali, Upper Volta
6. P. obovalis Tisserant	Central African Republic, Sudan
7. P. palustris Desv.	West Africa from Senegal to N. Cameroun, Sudan, and Central African Republic
8. P. scandens (Endl) Verdc.	Central and East Africa from Nigeria to Angola, Mauritius, Comoro Island, Madagascar. Cultivated in Jamaica, Brasil, India, Vietnam, Java, and New Guinea
9. P. tetragonolobus (L.) DC.	Cultivated in Africa, Asia, South Pacific islands, and South America

level of protein. The oil composition of seed is similar to
that of soybeans. Even the flowers, shoots and stems are
utilized.

Despite these qualities, only in Papua New Guinea and
Southeast Asia is the plant used extensively, and even there
it is a traditional poor man's crop. The winged bean's
growth potential, however, has been compared to that of the
soybean which has risen from neglect to one of the world's
premier protein sources in less than 60 years.

This review will summarize what is known about the
winged bean and assess the pros and cons of its potential as
an agricultural crop for the humid tropics and subtropics.

Taxonomy

The genus Psophocarpus Necker is composed of eight wild
species which are indigenous to Africa, and one cultigen.
Psophocarpus tetragonolobus (L.) DC., the cultivated winged
bean, is not known in the wild state but is grown throughout
Asia, including India, Sri Lanka, Burma, Thailand, Laos,
Vietnam, Malaysia, China, The Philippines, Indonesia, Papua
New Guinea, and some South Pacific islands. It has also been
reported from Sierra Leone, Ghana, Nigeria, Madagascar and
Mauritius, as well as South America. A monograph of the
genus is currently being prepared by B. Verdcourt of the
Royal Botanic Gardens, Kew, and will be published upon
completion in Kew Bulletin. Table 1 lists the nine
recognized species of Psophocarpus together with their known
distribution range (4).

The genus as a whole is characterized by climbing or
prostrate perennial herbs with pinnately trifoliolate or
unifoliolate leaves. Inflorescences are axillary, the blue
to purplish flowers solitary, fasciculate, or falsely
racemose. The pods are oblong and more or less distinctly
four-winged along the angles. The seeds are ovoid or
oblong-ellipsoid, with or without an aril (5).

Psophocarpus lukafuensis (De Wild.) Wilczek and P.
lancifolius Harms can be separated from the other members of
the genus by the presence of a characteristic style bearing
a ring of hairs below the apex. Both species possess
trifoliolate leaves with narrowly elliptic or rhombic-
lanceolate leaflets. Psophocarpus lancifolius is typically
pubescent to pilose with broader leaflets and larger flowers
than P. lukafuensis, and has been reported from swampy
grassland and forest edges at 1100 - 2550 m elevation (5).

Seven species are characterized by a style bearing hairs arranged laterally to form a tuft rather than a ring below the apex. Within this category, P. monophyllus Harms and P. lecomtei Tisserant are distinguished on the basis of their unifoliolate leaves. Psophocarpus monophyllus exhibits a trailing growth habit, broadly ovate or rounded elliptic leaflets and long inflorescence bracts. Psophocarpus lecomtei is typically prostrate with smaller, narrower rounded ovate or elliptic leaflets and shorter inflorescence bracts. The remaining five species possess trifoliolate leaves with broad leaflets. Psophocarpus obovalis Tisserant is a prostrate creeper with narrowly oblong or obovate leaflets. The large flowered P. grandiflorus Wilczek has been reported from upland bushland, forest and grasslands at an elevation of 1800 - 2100 m (5). Psophocarpus palustris Desv. was widely employed at one time to encompass popula-tions extending from east to west Africa, but two distinct species are currently recognized (6). Psophocarpus palustris is restricted in distribution to tropical West Africa, and is characterized morphologically by densely pubescent inflores-cence bracteoles which are shorter than the mature calyx, and short pods. Psophocarpus scandens (Endl.) Verdc. is separable on the basis of glabrous or only sparsely pubescent bracteoles equalling or exceeding the calyx in length and longer pods. It has a widespread distribution throughout central and east Africa (Table 1) and is cultivated to some extent in Central and South America as well as parts of Asia (4). In its natural habitat P. scandens occupies swamps, streamsides and ponds in forested areas at 0 - 810 m elevation, as well as disturbed grasslands and cultivated areas (5).

Cytology

Cytological investigations of Psophocarpus have resulted in an array of chromosome counts, leading to some confusion as to the exact complement of the genus. The following chromosome numbers have been reported for P. tetragonolobus: $2n=18$ (7, 8); $2n=$ca 18 (9); $2n=22$ (10); $2n=26$ (11). Studies of meiosis and mitosis in some accessions of P. tetragonol-obus carried out by the authors at the University of Illinois support a chromosome number of $2n=18$. A similar range has been recorded for P. palustris: $2n=$ca 18 (9); $2n=20$ (12); $2n=22$ (13). Cytological information on the remaining species is lacking at present.

Morphological Variation

Recent studies of P. tetragonolobus have revealed a large amount of morphological variation. Experimental work carried out on a large population of plants obtained from an extensive seed collection in Papua New Guinea has led to isolation of approximately 150 lines, each differing in morphological and growth characteristics (7, 14). Psophocarpus tetragonolobus is a twining perennial, usually grown as an annual, with large flowers and trifoliolate leaves. Leaflets are typically broadly rhomboid to ovate. Deltoid, ovate and lanceolate leaflet shapes have been noted in Javanese material (15). Stems range in color from green to deep purple. Flower colors are basically blue or purple, ranging from almost white to deep reddish purple in some lines. Pods also exhibit color variation, with a background of green, pink, or pale yellow, and various intensities of purple coloration in the wings resulting in wholly dark purple pods in some cases (7). The majority of pods show a rectangular shape in cross section with four wings extending from the angles. The wings themselves may be undulate, dentate, serrate, or lobed (15). At maturity the wings collapse and the pods become flattened. Other pod shapes include a semi-flattened and flat pod type, the latter compressed either along the sutures or along the sides of the pod wall (7). Pod surface may also be smooth or roughened. Seed color ranges from creamy white through brown to black (15, 16, 17), with shades of brown and tan being most common. A dark ring around the hilum and dark specks on the seed coat have been noted in some material (7).

Considerable variation has also been found in quantitative characters. Coefficients of variation calculated for material from Papua New Guinea ranged from 30% for mean seed weight and pod length to 83% for total seed weight per plant (7). Mature pod lengths have been reported to range from 12 - 70 cm in Java (15), 9 - 23 cm in Papua New Guinea (7) and 16 - 36 cm in Ghana (16), with an accompanying range in the number of seeds per pod and seed yield per plant. Tubers are produced by some winged bean lines, and were found in 16% of the Papua New Guinea Collection (7). Use of tubers appears to be restricted to Papua New Guinea and Burma (18) where plants have generally been selected either for high tuber or high pod yield.

Chemical Composition and Nutritive Value

Recent studies of P. tetragonolobus have revealed much chemical variation. Unfortunately, differences among the chemical values reported in the literature are confounded by years, locations, analytical and sampling methods and germplasm sources. Nevertheless, the values reported provide a reasonably good estimate of the chemical composition of winged bean seed, leaves, immature pods and tubers.

Table 2. The proximate composition of different parts of the winged bean expressed as g/100 g fresh weight. Compiled from Agcaoili (57), Anon (58), Bailey (59), Brown (40), Cerny et al. (21), Choo (27), Claydon (36), Jaffé and Korte (23), Padilla and Soliven (60), Platt (61), Watson (62).

Component	Immature pods	Seeds	Tubers	Leaves	Flowers
Water	75.9-92.0	8.5-14.0	51.3-67.8	77.7	84.2
Protein	1.9-2.9	32.4-41.9	4.7-20.0	5.7	5.6
Oil	0.2-0.3	13.1-19.1	0.1-0.4	1.1	0.9
Carbohydrate	3.1-3.8	25.2-32.0	27.2-30.5	3.0	3.0
Fiber	0.8-2.6	5.0-6.7	1.5-1.6		
Ash	0.4-1.9	3.6-5.8	0.9-1.7		

As shown in Table 2, the ranges in protein and oil contents in mature raw seed are from 32 to 42 and 13 to 19 g per 100 g seed (fresh weight basis), respectively. The protein and oil contents of most of the seed samples fall into the mid portion of the ranges. Therefore, winged bean seeds have about the same quantity of protein and oil as those of the soybean. Of particular interest are the protein rich tuberous roots (Table 2). The range in protein content of the tubers is from 4.7 to 20.0 g per 100 g, fresh weight basis. These protein values are from 5 to 15 times higher than the staple root crops of the humid tropics, taro, cassava, yams, potatoes and sweet potatoes (19). In addition to the protein rich seed and tubers, the leaves (20), flowers and immature pods also are reasonably good sources of protein (Table 2). Limited chemical data reveal that immature pods, seeds and tubers of the winged bean are rich in minerals and vitamins (Table 3).

A detailed amino acid composition of mature winged bean seed is shown in Table 4. The amino acid profile compares favorably with that of the soybean. However, like the soybean the sulfur containing amino acids, methionine and cystine, are the main limiting amino acids in winged bean seeds.

The fatty acid content of mature winged bean seed is shown in Table 5. The fatty acid profile is quite similar in quality to that of the soybean. Cerny et al. (21) reported that winged bean oil contains C 20:4 (parinaric acid), a potential antinutritional factor. However, Kleiman (22) using UV and GLC analytical procedures found that parinaric acid was not a component of winged bean oil. Most probably Cerny et al. (21) confused C 20:0 (arachidic) or C 20:1 (gadoleic ?) for parinaric.

Raw mature winged bean seed like the soybean and other legume seed contains various antinutritional or toxic factors such as amylase inhibitor (23), trypsin inhibitor (21) and hemagglutinin (24). Apparently, all of these seed components are inactivated or destroyed by moist-heat treatment or boiling in water.

The biological value of winged beans has been assessed in Ghana and Papua New Guinea with rats and human subjects. In rat experiments Cerny et al. (21) found that the protein efficiency ratio (PER) and net protein utilization (NPU) of winged beans were superior to that of peanuts. The PER and NPU values obtained with the winged bean were quite similar to that of the soybean (25). An experimental diet of 3 parts corn to 2 parts winged beans resulted in PER and NPU values

Table 3. The content of minerals and vitamins in different parts of the winged bean expressed as mg/100 g fresh weight. Compiled from Anon (63), Bailey (59), Brown (40), Cerny et aL (21), Church and Church (64), Claydon (36, 65), Jaffé and Korte (23), Platt (61), Watson (62).

Component	Immature Pods	Seeds	Tubers
Minerals:			
Calcium	25–236	204–340	25–40
Magnesium		57–255	23–64
Potassium	205	1010–1100	550
Sodium	3.1	20–64	33
Phosphorus	26–37	290–500	30
Iron	0.3–12	11–37	0.5–70.6
Manganese			3.0–5.0
Copper			0.6–6.8
Zinc			3.4–4.4
Sulfur			180–360
Vitamins:			
B-carotene I.U.*	595	330	
Tocopherols		126	
Thiamin	.1–.2	.2–1.4	
Riboflavin	.1	.2–.3	
Niacin	1.2	3.1	
Ascorbic Acid	19	0.7	26.2

*I.U., International Units.

Table 4. Amino acid composition (g amino acid/16 g N) of the protein of the winged bean (21) compared with that of the soybean (66).

Amino Acid	Winged Bean	Soybean
Arginine	6.5	7.6
Histidine	2.7	2.2
Lysine	8.0	6.0
Tyrosine	3.2	3.5
Trytophan	ND*	1.2
Phenylalanine	5.8	4.5
Cystine	1.6	1.3
Methionine	1.2	1.4
Serine	4.9	4.6
Threonine	4.3	3.7
Leucine	9.0	6.7
Isoleucine	4.9	4.4
Valine	4.9	4.6
Glutamic acid	15.3	18.4
Aspartic acid	11.5	10.4
Glycine	4.3	3.4
Alanine	4.3	3.6
Proline	6.9	5.3

*ND, not determined.

Table 5. Fatty acid composition of seed of winged bean (percentage composition by weight).

Fatty acid	Cerny et al. (21)	Kleiman (22)
14:0 Myristic	0.1	0.1
16:0 Palmitic	9.7	7.4
16:1 Palmitoleic	0.8	0.1
18:0 Stearic	5.7	2.8
18:1 Oleic	39.0	33.9
18:2 Linoleic	27,2	28.8
18:3 Linolenic	2.0	1.4
20:0 Arachidic	2.0	1.3
20:1 Gadoleic ?		4.0
20:2		0.1
20:4 Parinaric	2.5	none
22:0 Behenic	13,4	15.9
22:1		0.7
24:0		3.4

similar to that of skim milk. In Papua New Guinea, Jaffé
and Korte (23) observed that supplementation of winged bean
meal with 0.3% methionine increased PER values in rats nearly
to that of casein. Cerny and Addy (26) established that in
Ghana the winged bean is suitable for use as a milk substitute in the treatment of children for kwashiorkor. Results
from a preliminary study in Malaysia indicated that protein
from the winged bean after autoclaving could successfully
replace at least 50 percent of the protein from soybean meal
in chick rations (27).

Germplasm Resources

As of 4 years ago, there were no germplasm collections
of the winged bean nor were there any reliable sources of
seed for conducting extensive field, feeding or processing
experiments. This was a major obstacle to the investigation
of the potential of the crop (28). Today, Psophocarpus
germplasm collection and multiplication programs have been
established in Australia, Bangladesh, Colombia, Ghana, India,
Indonesia, Malaysia, Nigeria, Papua New Guinea, Philippines,
Sri Lanka, Thailand and the U.S. (2). In addition, The Asia
Foundation has contracted with the Florida Seed Foundation of
the University of Florida for large scale production, sale
and distribution of the cultivars 'Chimbu' and 'Tpt 1'. Seed
of these cultivars are now available from The Asia
Foundation.

Nodulation

Masefield (29, 30, 31) observed that the winged bean has
the capacity for forming exceptionally large and numerous
root nodules. In both Nigeria and in the Malayan Peninsula
he found that under field conditions the winged bean exceeded
all other legumes grown in those regions in the weight of
nodules produced. In Northwest Malaysia he found winged
beans producing 21 g of nodules per plant or the equivalent
of 672 kg of nodules per hectare. Masefield (28) hypoth-
esized that the high protein content of the seeds, pods,
leaves and tubers may be due to the crop's capacity for the
production of nodules. In a field experiment Iyer (32)
measured the mean fresh weight of nodules produced by legume
species in the tribes Galegeae, Genisteae, Hedysareae and
Phaseoleae. Of the species he studied only Crotalaria juncea
produced nodules with larger weights than the winged bean.
Thus Iyer confirmed the field observations of Masefield.

Cross inoculation relationships of the winged bean and
its rhizobium with other legumes and rhizobia were studied
by Elmes (33). He found that the winged bean belongs to the

cowpea cross inoculation group and is promiscuous with regard
to nodulation by rhizobium. At present, winged bean
inoculant is available from a private company and several
institutions maintain rhizobia strains for winged bean
research programs (1).

Utilization

Food

The winged bean is usually cultivated as an annual backyard or horticultural crop on a small scale throughout southeast Asia, and to a lesser extent in parts of Africa. At
present the plant is grown mainly for the tender young pods,
although all parts of the plant are edible and highly
nutritious. The use of different plant organs and methods of
their consumption vary from location to location. The
following uses have been reported and are listed according
to country.

Papua New Guinea. The winged bean forms an important
seasonal food in parts of Papua New Guinea, and is grown as
a field crop during the dry season in the Central and Western
Highlands. It may be second in importance only to the sweet
potato in the highland valleys (34). The green pods are the
most popular edible part, with tuber consumption confined to
relatively few tribes (35). Young pods and tubers are
traditionally cooked in "mumu" fashion, a method of steam
cooking either in a pit dug in the ground or on the surface
of the ground. Old pods are cooked in drums, the seeds and
surrounding mucilage being scraped out and eaten (36).
Leaves and flowers may be boiled or fried, eaten alone or as
a supplement to sweet potatoes or cooked bananas. Ripe seeds
may also be steamed, boiled, fried, or roasted within the
pods (34).

Indonesia. Young leaves, shoots and immature pods are
eaten raw as leafy vegetables, steamed, or cooked with other
vegetables to make side dishes and stews (15, 37). Ripe
seeds may be fried, roasted or boiled and eaten as a snack,
or fermented to make tempeh (37). It has been reported that
flower buds are consumed, and flowers used for coloring
certain dishes (38), but these uses have not been observed
recently (15). Swollen leaves, shoots and pods which have
become infected with false rust, Synchytrium psophocarpi
(Rac.) Gäumann, are regarded as a delicacy when boiled (37).
In Sulawesi and the Moluccas, pods may be fried with coconut
milk; in West Irian, green seeds may be boiled in a stew,
cooked in the fire, or steamed to use in vegetable side
dishes. Consumption of cooked tubers is rare or absent

throughout most of Indonesia, but has been recorded in Ambon and may still be practised there in some places (37).

The Philippines. Succulent, young winged bean pods may be sliced and cooked like French beans, added to stews, or served with dressing as a salad. Plants are often encouraged to grow against a native's hut for their attractive flowers (39), which may be used as a coloring for some dishes (40).

West Malaysia and Singapore. Young leaves are eaten raw or steamed; tender immature pods are cooked with chili peppers or eaten raw as a salad. Tubers are rarely eaten (37).

Thailand. Little is known concerning winged bean cultivation in Thailand, except that it appears to be widespread in the north and northeast provinces. Chopped immature pods are generally boiled, and roots consumed very rarely (37).

Burma. This appears to be the only country where winged bean tubers are used on a large scale, and winged beans may even be grown under irrigation (41). The tubers are slightly sweet with the firmness of an apple (18) and may be boiled or eaten raw. Claydon (37) reports that peeled small tubers, sometimes dipped in salt and oil, are highly popular as snacks. Young pods with unripe seeds are occasionally eaten.

Sri Lanka. Winged bean is grown as a home garden vegetable primarily for the immature pods, which may be chopped and cooked in coconut milk or tempered in coconut oil. They may also be eaten as a vegetable curry with rice (37). Tubers are eaten occasionally.

Bangladesh. Young pods are fried with fish or meat. The seeds are boiled and mixed with sugar, then eaten like nuts. Winged bean leaves and stems are fed to cows (1).

Mauritius. The winged bean has been recorded as cultivated and casually subspontaneous in Mauritius (42), but little is known of culinary practices except that the seeds are eaten (43).

Africa. Information concerning cultivation of winged bean in tropical Africa is also scanty. Young pods are mainly consumed (43), although young leaves and shoots may be eaten as a leafy vegetable, unripe seeds are used in soups, and ripe seeds may be roasted and eaten like peanuts (16). Immature tubers are eaten like potatoes.

Medicinal

Besides its many nutritional uses, the winged bean is valued for a variety of medicinal purposes. In Sumatra and West Java, an infusion of the leaves was traditionally used for eye and ear infections, leaves ground with aniseed and Alyxia stellata bark are used on boils, while seeds cleanse the blood and may be used in treatment of venereal disease (37). In central and East Java, Madura, and Bali, juices from the young shoots and leaves are used for teeth and morning dyspepsia, respectively. Pods are regarded as good for the blood and may be used in slimming diets, while the tuber boiled with sugar lumps forms a treatment for thrush (37). Pods are also thought to be good for the blood in Sulawesi and the Moluccas (37). In Malaysia, winged bean leaves are boiled together with the leaves and tubers of several other plants to make a lotion used in curing small-pox (44). The tuber may be used in the Shan States of Burma as a poultice for the head and neck in the treatment of vertigo (18). In Sri Lanka, the pod is regarded as useful for diabetes (37). A decoction of leaves and roots of P. scandens has been used in Tanzania to reduce lactation and to treat sores (4).

Other

It has been suggested that the winged bean has potential as a cover crop or fallow restorative crop, on account of its highly effective nodulation system and resulting tolerance of poor soils (31). It has been tried as a green manure in Java, but reportedly was unsatisfactory (18) possibly due to the difficulty of incorporating the mass of foliage into the soil. Psophocarpus tetragonolobus is sometimes grown as a cover crop in banana plantations in southeast Asia (39). The feasibility of using it as a multirole second crop in rubber plantations and smallholdings is being assessed in Malaysia and Sri Lanka (1, 45). The smaller leaved P. palustris has been reported to grow well as an oil palm cover crop on some soils in Sumatra, but poorly on others in Malaysia, while in West New Britain it formed a good cover after rather a slow start (46). Psophocarpus palustris also gave promising results as a ground cover in Sri Lanka (47) showing no signs of dying back even after two years.

Psophocarpus scandens [as P. palustris, see (6)] has been found under cultivation to some extent, being used as a food source in the same way as P. tetragonolobus. It has been stated that the ripe seed are used in Indonesia (18), and young pods, shoots, and leaves are eaten in Madagascar and Congo (43). The root of P. lecomtei is reportedly used

to kill fish (48), which may suggest the presence of a
rotenone-like poison in the rootstock (4). Psophocarpus
grandiflorus has also been mentioned as a fiber plant (48)
but no further details were given.

Diseases

The winged bean hosts few pathogens of current economic
significance. Fungal diseases found both on cultivated
plants and in experimental plots include powdery mildew (49),
leaf spot (50), bean anthracnose (51), and false rust (50).
Psophocarpus tetragonolobus was found to be susceptible to
cowpea mosaic in Trinidad (52) and immune to bean fly in the
Philippines (53). Perhaps the most damaging and widespread
pest is the root-knot nematode found throughout the highlands
and lowlands of Papua New Guinea and New Britain (50). The
apparent absence of serious pests and diseases is probably
due to the fact that the crop is not planted as a monoculture
over large areas. Populations are usually small, well
spaced, and heterogeneous, all of which would serve to
prevent pathogen buildup. Under a more intensive farming
system the crop is likely to suffer more from disease than
at present (51).

Production

It should be emphasized that one of the major drawbacks
to the investigation of the agronomic potential of the winged
bean continues to be that of seed supply. Accordingly, there
have been no large-scale experiments designed to evaluate
and compare agronomic properties of winged beans from many
different lines.

The winged bean is a twining vigorous perennial usually
grown as an annual. It thrives best in the tropics under
conditions of high rainfall, in excess of 250 cm, and thus is
usually planted at the onset of the wet season. It may be
grown in the dry season as long as there is still adequate
rainfall, or supplemental irrigation; in Papua New Ginea,
cultivation of the plant during the dry season minimizes
damage and loss due to pests and diseases (34). The winged
bean is able to grow on relatively poor soils owing to its
extensive nodulation system, which develops even in freshly
cleared forest soils where Psophocarpus has not been planted
previously (28).

Since the plant is a vine it must either be staked up or
grown along a fencerow for support. Large pure stands of the
winged bean are relatively uncommon, the plant usually being
grown as a backyard vegetable supply in conjunction with

other food crops. Cultivation practices vary depending on whether the plant is used as a source of green vegetables or tubers. Those grown for tuber production are either supported on short stakes or allowed to ramble on the ground; young shoots, flower buds and pods are picked off periodically to encourage tuber formation. In the highlands of Papua New Guinea a section of a mixed garden may be planted to winged bean, with certain areas in the section being set aside for tuber production (17).

The time taken for the winged bean to flower after sowing depends upon the season and location, as the plant is daylength sensitive. Time from planting to flowering may vary from 4 to 36 weeks (37). Young green pods are ready for harvesting about 2 to 3 weeks after fertilization, and flowers and pods continue to be produced over a period of several weeks. The winged bean appears to be mainly self-pollinated, although it has been reported in Ghana that exclusion of bees and other pollinating agents reduces yield significantly (54). Mature fruits and ripe seeds can be picked 3 - 7 months after planting, while tubers are harvested 5 - 12 months after planting (37).

Few trials have been carried out to evaluate yield potential of the various parts of the winged bean that are utilized. Those experimental data which are available come mainly from small plot trials, and projected yields in terms of kg/ha are obtained by extrapolation. Direct comparisons between yields reported in the literature are not necessarily reliable owing to differences in plant populations per area, information which is not always supplied. Few figures are available for yield of green pods; values up to 35,526 (27) and 34,710 (55) kg/ha fresh weight have been obtained. Projected seed yields of 1409 (16), 1946 (7), and 4590 (27) kg/ha have been reported. Tuber yields vary widely, the following ranges having been recorded: 1350 (7), 2246 - 4980 (27), and 5533 - 11,754 (34) kg/ha fresh weight.

As a kitchen garden vegetable the winged bean has several advantages. It is mainly cultivated as a source of young green pods, and the indeterminate growth habit ensures that production of young shoots, flowers and pods continues over an extended period of time. High yielding plants of this nature can thus provide a protein-rich supplement to the diet of a family for several months, especially since all parts of the plant are edible and high in protein.

The crop in its present form, however, does not lend itself to commercial production unless labor is cheap and readily available. Staking plants over large areas becomes prohibitive, and simultaneous ripening of fruits is necessary for mechanical harvesting. Production of seed on a large scale would become feasible with the development of a free standing, determinate cultivar, which could be harvested mechanically in the same way as the soybean, and experimentation with this aim in mind is being carried out in Papua New Guinea (56). Agcaoili (57) noted that owing to the similarity between soybean and winged bean seed, whatever use was currently being made of the soybean in China, Japan and the Philippines could also be made of the winged bean. Winged bean seeds are already fermented to produce tempeh in parts of Indonesia (37) but this usage appears to be rare at present. Research on production of winged bean curd and milk in Malaysia and Thailand have so far given promising results (1). Seed production on a commercial scale would greatly increase the availability of such products in the protein-deficient tropics.

The pronounced nodulating ability of the winged bean makes it a useful component of crop rotations for enriching the soil, and possibly as a cover crop in plantations.

The winged bean's potential as a backyard vegetable crop capable of supplying protein locally should be exploited to the fullest, particularly in those high rainfall areas of the tropics where commercial production is difficult owing to climate or terrain and where the traditional staple food crops are highly protein deficient.

For the latest information concerning international efforts on behalf of the winged bean, please contact Mr. Louis Lazaroff, The Asia Foundation, P. O. Box 3228, San Francisco, CA 94119, U.S.A.

Note

This paper is a contribution from the Department of Agronomy, University of Illinois at Urbana-Champaign, Urbana, IL 61801. Research supported in part by the Illinois Agricultural Experimental Station and the Asia Foundation.

References and Notes

1. J.W. Levy, The Winged Bean Flyer (Department of Agronomy, University of Illinois in cooperation with the Steering Committee of the Winged Bean, 1977), vol. 1, no. 1.
2. _____, ibid.vol. 1, no. 2 (1977).
3. National Academy of Sciences, The Winged Bean: a High-Protein Crop for the Tropics (Washington, D.C., 1975).
4. B. Verdcourt, personal communication.
5. _____, in Flora of Tropical East Africa, E. Milne-Redhead and R.M. Polhill, Eds. (Crown Agents for Oversea Governments and Administrations, London, 1971), pt. 83, pp. 602-606.
6. _____, Taxon 17, 537 (1968).
7. T.N. Khan, Euphytica 25, 693 (1976).
8. P. Tixier, Rev. Cytol. Biol. Vég. 28, 133 (1965).
9. J.A. Lackey, dissertation, Iowa State University, Ames (1977).
10. N.V. Thuan, Rev. Gén. Bot. 82, 157 (1975).
11. D.A. Ramirez, Philipp. Agric. 43, 533 (1960).
12. J.A. Frahm-Leliveld, Acta Bot. Neerl. 9, 327 (1960).
13. J. Miège, in Index to Plant Chromosome Numbers for 1963, M.S. Cave, Ed. (University of North Carolina Press, Chapel Hill, 1964), p. 314.
14. T.N. Khan, The Situation of the Germplasm Resources of the Winged Bean (mimeo report of Third SABRAO Congress, Canberra, Australia, 1977).
15. S. Sastrapradja and S.H. Aminah Lubis, in South East Asian Plant Genetic Resources, J.T. Williams, C.H. Lamoureux, N. Wulijarni-Soetjipto, Eds. (International Board for Plant Genetic Resources, Bogor, Indonesia, 1975), pp. 147-151.
16. F. Pospisil, S.K. Karikari, E. Boemah-Mensah, World Crops 23, 260, 1971).
17. J. Powell, Science in New Guinea 2, 48 (1974).
18. I.H. Burkill, Dictionary of the Economic Products of the Malay Peninsula (Crown Agents for the Colonies, London, 1935), vol. 2.
19. W.W. Leung, Food Composition Table for Use in Latin America (The Institute of Nutrition of Central America and Panama, Guatemala City and the Interdepartmental Committee on Nutrition for National Defense, National Institutes of Health, Bethesda, Maryland, 1961).
20. Y.D.A. Senanayake and V.A.D. Sumanasinghe, J. Natl. Agric. Soc. Ceylon 13, 119 (1976).
21. K. Cerny, M. Kordylas, F. Pospisil, O. Svabensky, B. Zajic, Brit. J. Nutr. 26, 293 (1971).
22. R. Kleiman, personal communication.
23. W.G. Jaffé and R. Korte, Nutr. Rep. Int. 14, 449 (1976).
24. D.O. Renkonen, Ann. Med. Exp. Biol. Fenniae 26, 66 (1948).

25. I.E. Liener, in Soybeans: Chemistry and Technology, A.K. Smith and S.J. Circle, Eds. (AVI Publishing Company, Westport, Connecticut, 1972), pp. 203-277.
26. K. Cerny and H.A. Addy, Brit. J. Nutr. 29, 105 (1973).
27. W.K. Choo, The Potential for Four-angled Bean [Psophocarpus tetragonolobus (L.) DC.] in Malaysia to Increase Food Supply (Umaga/Faum Food Conference, Universiti Pertanian Malaysia, 1975).
28. G.B. Masefield, Field Crop Abstr. 26, 157 (1973).
29. _____, Empire J. Exp. Agric. 20, 175 (1952).
30. _____, ibid. 25, 137 (1957).
31. _____, Trop. Agric. (Trinidad) 38, 225 (1961).
32. N.R. Iyer, Pl. & Soil 44, 451 (1976).
33. R.P.T. Elmes, Papua New Guinea Agric. J. 27, 53 (1976),
34. T.N. Khan, J.C. Bohn, R.A. Stephenson, World Crops & Livestock 29, 208 (1977).
35. _____, in South East Asian Plant Genetic Resources, J.T. Williams, C.H. Lamoureux, N. Wulijarni-Soetjipto, Eds. (International Board for Plant Genetic Resources, Bogor, Indonesia, 1975), pp. 152-156.
36. A. Claydon, Science in New Guinea 3, 103 (1975).
37. _____, Winged Bean in Southern Asia, summary of information collected on study leave January-February 1977 (mimeo report, Department of Chemistry, University of Papua New Guinea, 1977).
38. O. Degener, J. New York Bot. Gard. 46, 110 (1945).
39. G.A.C. Herklots, Vegetables in South-East Asia (George Allen and Unwin Ltd., London, 1972).
40. W.H. Brown, Useful Plants of the Philippines (Technical Bulletin No. 10, Bureau of Prints, Manila, 1954), vol. 2.
41. J.W. Purseglove, Tropical Crops: Dicotyledons (John Wiley and Sons, Inc., New York, 1968), vol. 1.
42. J.G. Baker, Flora of Mauritius and the Seychelles (L. Reeve and Co., London, 1877).
43. C. Jardin, List of Foods Used in Africa (FAO, Rome, 1967).
44. I.H. Burkill and M. Haniff, Gard. Bull. Straits Settlem. 6, 165 (1930).
45. Y.D.A. Senanayake, Bull. Rubber Res. Inst. Sri Lanka 11, 16 (1976).
46. N.J. Mendham, Papua New Guinea Agric. J. 22, 250 (1971).
47. W.R.C. Paul, Trop. Agric. (Ceylon) 107, 225 (1951).
48. R. Wilczek, in Flore du Congo Belge et du Ruanda-Urundi: Spermatophytes (Publications de l'Institut National pour l'Étude Agronomique du Congo Belge, Bruxelles, 1954), vol. 6, pp. 280-288.
49. T.V. Price, Pl. Dis. Reporter 61, 384 (1977).
50. _____, Austral. Pl. Pathol. Newslett. 5, 1 (1976).
51. P.J. Keane, Science in New Guinea 2, 112 (1974).

52. W.T. Dale, Ann. Appl. Biol. 36, 327 (1949).
53. O.Y.F. Quesales, Philipp. Agric. 7, 2 (1918).
54. S.K. Karikari, Ghana J. Agric. Sci. 5, 235 (1972).
55. F. Jimenez, Vegetables for the Hot, Humid Tropics 1, 27 (1976).
56. T.N. Khan and A. Claydon, in Evaluation of Seed Protein Alterations by Mutation Breeding (International Atomic Energy Agency, Vienna, 1976), pp. 209-210.
57. F. Agcaoili, Philipp. J. Sci. 40, 513 (1929).
58. Anon, Grain Legume Improvement Program (International Institute of Tropical Agriculture, Ibadan, Nigeria, 1973).
59. K.V. Bailey, Trop. Geogr. Med. 20, 141 (1968).
60. S.P. Padilla and F.A. Soliven, Philipp. Agric. 22, 408 (1933).
61. B.S. Platt, Tables of Representative Values of Foods Commonly Used in Tropical Countries (Medical Research Council Special Report Series No. 302, HMSO, London, 1962).
62. J.D. Watson, Ghana J. Agric. Sci. 4, 95 (1923).
63. Anon, The Wealth of India (Publication and Information Directorate, New Delhi, 1969), vol. 3.
64. C.F. Church and H.N. Church, Food Values of Portions Commonly Used (J. Lippincott, Philadelphia, ed. 11, 1970).
65. A. Claydon, An Investigation into the Storage of Winged Bean Roots, Psophocarpus tetragonolobus (L.) DC. (mimeo report of 10th Waigani Seminar, Lae, Papua New Guinea, 1976).
66. J.J. Rackis, R.L. Anderson, H.A. Sasame, A.K. Smith, C.H. van Etten, J. Agric. Food Chem. 9, 404 (1961).

3

Amaranth

Gentle Giant of the Past and Future

Laurie B. Feine, Richard R. Harwood, C.S. Kauffman and
Joseph P. Senft

Abstract

The many species of the family *Amaranthaceae* form an
extremely diverse group of plants with worldwide distribution.
Several species with a history of thousands of years of cultivation have been used as vegetable or as grain producing
plants. Grain amaranth reached a peak of popularity as a
staple crop during the Mayan and Aztec periods in Central
America. The grain has high nutritional value, containing
12-15% protein with a high lysine level. The young leaves
of types selected for vegetable use are similar to spinach
and other crops normally used as cooked greens. These types
are widely grown in Asia. Current crop improvement research
in the U. S. is focused on worldwide germplasm collection,
genetic studies and varietal selection for both home garden
and commercial use. The relative ease of intercrossing
among many of the domesticated species and between domesticated and wild species is providing an extremely broad
genetic base for early crop improvement work. The yield
potential for both grain and vegetable types appears similar
to that of currently used vegetable and cereal crops.

Introduction

Amaranth has received increased scientific attention
in recent years as we have searched for "bypassed" crops
having potential for broadening man's food base. Its at-
tractiveness as a crop for the future stems from extremely
broad climatic adaptability, a rich germplasm resource
with genetic variability exceeding that of most commercial
crops, high protein-quality in its grain, good grain yield
potential and its widespread use in Asia as a cooked leafy
vegetable (1). Except for vegetable use, amaranth culture
is confined to small pockets in the Americas and Asia, with
little movement of seed and only sporadic research attention.

The domesticated vegetable and grain types are grown mostly
in the tropics and subtropics, but appear to grow equally
well as warm season temperate zone crops. The grain types
are primarily found in semi-arid, seasonally wet areas. It
thus represents a crop that has, for the most part, been
neglected during the period of rapid crop advances of the
last two centuries.

History and Use

Amaranths are one of the oldest food crops in the new
world (2). Archeological digs in the Tehuacan Valley have
found cultivated amaranths dating from 6700 to 5000 B.C. (3).
The long association of man and amaranths is probably due to
the ability of these plants to adapt readily to new environments created by man. Their competitive ability permitted
culture with minimal crop management. They are adapted to
open sun and disturbed habitat. It thus became an easy crop
for early people to cultivate and domesticate. Distinctive
grain and vegetable species were developed, each involving
separate cultural techniques and uses.

Amaranth as a domesticated grain crop reached a high
point during the reign of the Aztec empire. Its use virtually equalled that of corn and it became an important religious symbol. After the overthrow of the Aztec empire by
Cortes, the cultivation of amaranth was suppressed because
of its pagan symbolism. Since that time, cultivation in the
Americas has declined to only small pockets scattered
throughout Central and South America. The main areas of
grain cultivation are now in Asia, centered in Nepal and
certain Indian states (4).

Amaranth is more widely used as a potherb. Cultivated
for vegetable use throughout the tropics and eastern Asia,
it is considered one of the best tropical greens. Its
mild flavor, good yields, ability to grow in hot weather,
and high nutritive value have made it a popular vegetable (5).

The history of amaranth has been described as "neglected
and obscure" (6). Most research that has been conducted to
date on this genus has been related primarily to anthropology,
botany and plant physiology. The plant has the C_4 pathway
with its high photosynthetic efficiency (7). It is one of
the few dicotyledons with potential for becoming a grain crop.
Much of the literature leaves the reader with the impression
that the genus, *Amaranthus*, although interesting from a
scientific point of view, has little application to modern
agricultural practice.

Taxonomy and Genetics

The family *Amaranthaceae* (Dicotyledons, order Caryo-
phyllales) is composed of 60 genera and about 800 species.
They are all annual herbaceous plants with many species
having tropical origin, but with adaptation to temperate
climates as well. Taxonomic classification has been difficult. Sauer (8) attributes the taxonomic problems to

> *"the hopeless attempts to recognize by pigmen-
> tation which segregates within populations,
> and growth which is extremely plastic under
> different day lengths and other environmental
> variables."*

Sauer's·classification relies on species identification
through constant characters, particularly shapes and proportions of pistillate flower parts.

The genus *Amaranthus* has been classified according
to floral structure by various researchers (4,8,9,10). Two
sections have been listed within the genus *Amaranthus:*
Section *Amaranthus* and Section *Blitopsis.*

The section *Amaranthus* includes those species which
are usually considered to be grain types, including the
dye amaranths, most of the domesticated ornamentals, the
majority of the vegetable types, and most of the common
weeds (5,8). This includes the species *A. cruentus, A.
caudautus, A. hypochondriacus* and *A. edulis.* The group
normally has a compound terminal inflorescence (4) that
is indeterminate in all species except *A. edulis* (10).
Flowers are usually pentamerous with a dehiscent utricle
that is circumscissle (10).

The grain amaranths are further distinguished by relatively short and weak bracts (which may result from artificial
selection against prickliness). Seed yield, but not seed
size, has been selected for by increased plant and inflorescence size. Pale seed, which is found only in the grain
types, is a result of the universal preference for light
colored grains (8).

The section *Blitopsis* has axillary, determinate flowers
and if there is a terminal inflorescence, it is very small.
Flowers are usually bi- or tri-merous with an irregular
dehiscent utricle (10). The section *Blitopsis* includes
the vegetable species *A. gangeticus, A. tricolor* and *A.
blitum* (5).

The basic floral structure for both sections is a dichasial cyme commonly called a glomerule. An initial staminate flower is followed by an indefinite number of pistillate flowers. The glomerules are on a leafless axis and form a complex panicle which is technically called a thryse. Prior to stamen exertion, the pistils within a given glomerule are receptive to pollen (4).

All amaranths exhibit varying phenotypes and are adapted to a variety of climatic conditions. The most common leaf shape is elliptical with an acute tip and a cuneate base. Leaf size varies greatly between and within species. Plant color also varies greatly from dark green to magenta with a multitude of intermediates and combinations.. The characteristics of varying phenotypes, although impossible to use for taxonomic purposes, are one of the unique features of this family, having been a major factor in contributing to its cosmopolitan nature.

The genus *Amaranthus* has created widespread interest in the scientific community (4,11,12,13) because of its movement around the world during the 16th and 17th centuries, and its seeming ability to hybridize freely within the genus. Its advantages as a tool for genetics research are that it can be grown in a small space due to its plastic morphology, it has the ability to produce as many as six generations per year and it is capable of both self and cross-pollination with no marked inbreeding depression or heterosis (9).

The interest generated by these factors has provided a relatively rich literature on amaranth genetics. Three recurring subject areas have been explored at length: 1) Interspecific hybridization 2) Sex expression and 3) Induced polyploidy.

The use of chromosome counts has not been especially helpful in defining the different species. Diploid chromosome counts have been reported as either 32 or 34 (4,14,15), often within the same species. Sauer (8) notes that the two different diploid chromosome counts within a species is not necessarily an indication of different species or an indication of incompatability in crossing.

Hybridization has been reported between many species (16,17,18). The reported failures to recombine have been in crosses between species of the section *Amaranthus* and species of the section *Blitopsis*.

Interspecific crosses within the section *Amaranthus*

have been reported (12,14,19,20,21). In some cases, the F_1 hybrids are sterile (11,22,23) although reports of successful F_2 and subsequent generations have been noted (12, 21). Naturally-occurring, interspecific hybrids have also been found (11,90). At least half of the species of the genus have been involved in hybridization (14).

By definition, a species is isolated from other species by barriers of sterility or reproductive incapacity (24). The naturally-occurring, interspecific hybridization within the genus is one of the major factors which has caused the within-species variation and taxonomic complexity of *Amaranthus* (14).

The amaranth species that are commonly used as grains and vegetables are all monoecious. This causes problems for the plant breeder and has prompted research on the use of dioecious species.

In 1938, Murray (20) completed successful "intergeneric" crosses of monoecious species and dioecious members of the genus *Acnida*, a member of the family *Amaranthaceae*. Segregation for sex expression was noted among the progeny. No F_2's were made. In 1955, Sauer (23) reclassified *Acnida* as part of the genus *Amaranthus*. The three most common dioecious species are *A. tuberculatus*, *A. tamariscinus* and *A. arenicola*. All dioecious species are native to the heartlands of North America and have had many contacts with both *A. retroflexus* and *A. hybridus*. Naturally occurring F_1 hybrids have been observed. However, pollen investigations have all indicated that pollen is sterile. Sauer (11) notes that all crosses between monoecious and dioecious species provide sterile hybrids, thus, eliminating any possibility of using dioecious species for plant breeding.

Amaranth seeds are very small, with 1,000 seeds per gram being common. Although selections have been made over the years for pale grain color (vs. black in the wild species) and more seeds per plant, there has been apparently little selection for larger seed size (25).

Pal and Khoshoo (25) created tetraploid lines by treating seeds of *A. edulis* with colchicine. The tetraploids were shorter, sturdier and non-lodging in comparison to their diploid progenitors. Furthermore, tetraploid lines exhibited a 2.5 fold increase in seed weight.

In 1977, additional reports were made on colchicine-induced polyploidy in *A. edulis*, *A. hypochondriacus* and *A. caudatus* (25). Progeny were stockier, with smaller,

broader and thicker leaves than their diploid parents. The
monoecious flowering habit was disturbed. Predominately
male plants were found among *A. edulis* and *A. hypochondriacus* and some completely female plants were found in *A.
caudatus*.

Seed size was significantly increased in all tetraploids.
In spite of the imbalance in sex determination, pollen fertility remained reasonably high (greater than 82%).

Misra et al., (26) report that polyploidy does not
adversely affect the amino acid spectrum. In fact, an increase in lysine was noted in the tetraploid of *A. edulis*.

Nutritional Qualities

Amaranth constitutes an important part of the diet in
areas of South America, Africa and Asia (4,5). It is used
as both grain and potherb and in some instances supplies
a substantial portion of the protein, minerals and vitamins
in the diet (61). The grain protein is unusual because its
amino acid compliment is very similar to the optimum balance
required in the human diet. The lysine content is especially
high in comparison to that found among the more common
grains. The nutritional qualities of amaranth thus may be
important to its development as a food source for low income
people of third world countries as well as to its use as an
alternative crop in the industrialized world.

Elias (27) has compared the nutrient composition of
amaranth seeds with an average for that of cereals in
general in Table 1.

TABLE 1.   Comparison of Amaranth Food Value with that of
Other Cereals

	Food Constituent (grams/100 grams dry weight)							
	Protein	Fat	Carbohydt.	Fiber	Ash	Ca	P	Fe
Cereals	11.0	2.7	73.0	2.1	1.7	.03	.33	.0034
Amaranth	14.5	7.5	60.4	7.5	2.9	.37	.48	.0034

Amaranth grain has a relatively high protein, fat and mineral content. Its fiber content is 3-4 fold that of common grains. Its amino acid composition makes it a more complete protein source. Amino acid chemical scores for *A. hypochondriacus* grain and *A. retroflexus* grain were 75 and 87 respectively in comparison to maize-44, wheat-57, sorghum-48 and barley-62 (27). Schmidt (28) reported protein levels of 12.6-15.6% for *A. hypochondriacus*. The level of mineral constituents in amaranth were higher than those in barley, triticale and wheat grown under similar conditions. Calcium was 3-4 fold greater, magnesium was 2 fold greater, iron was 5 fold greater and zinc was 1.5 fold greater. Starch from *A. cruentus* and *A. leucosperma* is similar to the waxy varieties of cereal grains rather than the starches found in the more common grains based on their color reaction with iodine-potassium iodine and iodine-sorptive capacity (29,30). Some physical properties of the starch from *A. retroflexus* have been reported (31). An early assessment of the potential for amaranth as a new source of grain based on protein quality alone concluded that its potential was limited in comparison to that of other species (32). Amaranth grain is clearly, however, a balanced food.

Vegetable Amaranth

Leaves of probably all *Amaranthus* species are edible and many appear regularly in diets of several societies (5,33). Vegetable types have been genetically selected and are available from commercial seed companies.

Nutritional data for a number of vegetable amaranth species show the following in comparison to spinach (27) in Table 2.

TABLE 2   Comparison of Amaranth and Spinach Food Value

| | Grams per 100 grams fresh wt. | | | | |
	Protein	Fat	Carbohydt.	Fiber	Ash
Amaranth	1.6-5.6	0.1-0.8	4.5-17.9	1.0-2.1	1.0-2.5
Spinach	3.2	0.3	4.3	0.6	1.5

Figure 1.   Vegetable amaranth variety recently introduced from
The People's Republic of China growing well in the
United States.   --Rodale Press Photo

Protein levels (dry weight basis) of 11.3% to 27.7% for *Amaranthus hybridus* (34) 26.7% and 27.8% for *A. lividus* and *A. hybridus,* respectively (36) and 29.7% for *A. caudatus* (36) have been reported.

The amino acid composition of *Amaranthus hybridus* as compared to protein from other sources shows a chemical score of 71 in comparison to 68 for spinach, 74 for soybean, 67 for rice and 53 for wheat (37). Amaranth protein may provide as much as 25% of the daily protein intake in one African society (5).

A further potential of vegetable amaranth protein may be realized through the development of leaf protein concentrates. Amaranth can produce up to 5,000 kg of protein/ha. It had the highest level of extractable protein among 24 plant species studied (38).

A comparison of the mineral content between amaranth and spinach is given in Table 3 (27).

TABLE 3   Mineral Content of Amaranth and Spinach

		(milligrams/100 grams wet wt.)			
	Ca	P	Fe	Na	K
Amaranth	146-476	45-123	2.2-16.0	4-115	411-575
Spinach	93	51	3.1	71	470

Levels of micronutrients such as zinc, manganese and copper have been reported (39,40) for *A. gangeticus* and *A. cruentus.*

Amaranth, like many other rapid growth plants, requires and absorbs large amounts of nitrate which is necessary for protein synthesis. Under certain conditions, nitrate can accumulate to levels which constitute several percent of plant dry matter (41,42,43). Absolute levels in amaranth are about the same as those in spinach. Nitrates can be converted to nitrites in the digestive tract and thus may become precursors for nitrosamines. White (44) estimates that about 86% of the average daily intake of nitrate is derived from vegetable sources. The extent to which dietary nitrate contributes to the production of nitrosamines, however, is dependent upon many factors which require further investigation before conclusions can be drawn regarding the specific relationships between nitrate and nitrosamine production (45). In any event, it does seem to be desirable to develop vegetable varieties which have low nitrate levels. This is one quality which should be included in future amaranth selection (46).

Nitrate is absorbed from the digestive tract and excreted in the urine by humans and other monogastric animals. Cattle, however, can convert large amounts of nitrate to nitrite through their rumen. This can lead to methemoglobinemia, cyanosis and death (41,47). Development of amaranth as a forage crop will thus require consideration of nitrate accumulation characteristics.

Green leafy vegetables also frequently contain large quantities of oxalate. This is characteristic of amaranth as well. Oxalate is an end product of metabolic processes in plants, and in many plants, accumulates as the plant becomes older (48). It is also an end product in animal metabolism and is well known in kidney pathologies such as the formation of kidney stones (48,49). Oxalate binds divalent cations, particularly calcium, and in the digestive tract may make these unavailable for absorption (50, 51,52). Despite the strong binding of calcium to oxalate, amaranth apparently can be a dietary source of calcium (53,54,55). A study of the effects of oxalic acid on availability of zinc from spinach leaves and zinc sulfate in rats (56) demonstrated that oxalic acid levels did not affect zinc absorption, contrary to expectations. Further study is needed to determine relationships between intestinal oxalate levels and divalent cation absorption. Oxalate levels in plants are affected by soil fertility, showing an increase with increased fertilization (57). There is potential for selecting amaranth varieties raised under specific growing conditions for optimal growth, but minimum accumulation of oxalates.

Vitamins

Elias (27) has compared the vitamin levels in various amaranth varieties with those of other green vegetables. The range for these amaranth varieties in comparison with spinach is given in Table 4.

TABLE 4   Comparison of Amaranth and Spinach Vitamin Levels

	Amounts per 100 grams fresh wt.				
	Vit. A I.U.	Thiamine mg.	Riboflavin mg.	Niacin mg.	Ascorbic Acid mg.
Amaranth	65-7715	.01-.08	.14-.42	0.3-1.8	12-120
Spinach	8100	.1	.2	.6	51

Green leafy vegetables become an important source of vitamins where preformed vitamins are too expensive for the poor. Amaranth as a source of vitamin A in preschool children has been found to be a readily utilizable source (58).

Green leafy vegetables are known to be manufacturers of a wide variety of products of pharmacologic significance. Amaranth is no exception. Betacyanes, alkaloids such as betaine, cyanogenic compounds, saponins, sesquiterpines, polyphenols all have been reported to be present in various members of the *Amaranthaceae* (47,59,60,61). Thus, selection for varieties which have low levels of potentially anti-nutritive substances is another consideration in the development of economically useful amaranth species. For example, studies similar to those reported for two alfalfa varieties which contain different levels of saponins will be useful (62). Thus, while genetic diversity in the genus is great, care should be taken during crop improvement to insure that the nutritional quality of selected types remains high.

Culture of Vegetable Amaranth

While vegetable amaranth is included in most listings of tropical vegetables (63) and is cultivated or gathered in many tropical countries (especially in Southeast Asia) (64), there are few references that include more than generalities about its culture. This may be an indication both of its ease of cultivation and the fact that, because of wide adaptability, the optimal conditions for maximum yields are not known. Varietal selection for vegetable types by growers has, over the years, resulted in the emergence of cultivars with leaves and stems of high eating quality. Most have small, black seeds.

General recommendations usually suggest well-prepared seed beds, a fertile soil that is well drained (while still retaining moisture), and an irrigation source. Spacing depends on harvest method used, e.g., harvest by pulling (clear cut) or multiple harvests (repeated cutting with time for regrowth between) (65). Cultivations may take two forms (64).

1. Dense sowing in rows 20 cm. apart in which the plant is pulled after some weeks.
2. Wide sowing with transplanting at a density of 30 cm. x 50 cm. with each plant ratooned several times.

The first method requires a large amount of seed for each sowing. It is assumed that there will be multiple sowings during the growing season requiring an even larger amount of seed. In Africa, all beds are established by transplanting from a seedbed in order to save seed and reduce the amount of labor which is expended for thinning (5).

Transplanting has been shown by Mohideen and Rajapal (66) to delay flowering and increase the total duration of the vegetative phase.

Seed germination is dependent on warm weather and adequate moisture. A number of germination physiology studies have been conducted for weedy types. The optimal temperature for germination of *A. retroflexus* has been reported to be 35°C (67), with a range of 30°-35° to be acceptable for a good germination percentage (68,69). Germination inhibition has been noted in vegetable types at a temperature of 45°C (70). Weedy species which have undergone natural selection in their northward spread have larger seeds and an ability to germinate at lower temperatures (67). Problems with stand establishment may occur when tropical vegetable types are sown too early in the season in temperate climates.

Vegetable amaranth yields have been reported as high as 40 MT/ha (5). Fertilization, especially nitrogen applications, seems to be one of the major factors influencing yield (5). Fertilizer NPK as 10-10-20 at a rate of 400 kg/ha at time of plowing produced the highest yields in a fertility trial (5). Deutsch (63) reported that "fresh and dry weight yields increase linearly with nitrogen application from 0 to 200 kg/ha."

Insect and disease problems can seriously affect yield.

Damping off from *Pythium* and *Rhizoctonia* is most serious
during humid weather, particularly during a monsoon rainy
season. Seed beds (especially those established during
humid weather) should be well-drained and sunny. Amaranth
does not grow well during long periods of cloudy, wet
weather. It is not tolerant of shade.

Another disease, *Choanephora cucurbitarum,* causes
wet-rot of leaves and young stalks. Manuring seems to
eliminate some of the problems. Various fungicides have
also been used successfully (5).

Insect damage can be a more serious problem in vegetable
amaranth than in grains because the vegetative parts are
harvested when the plant is very young. The major insects
are larvae of the order Lepidoptera, leaf hoppers, grass-
hoppers and leaf feeding beetles.

Since only the tender foliage and stems are harvested
and the plant grows rapidly, the time for harvest is
relatively short. Yields are greatly reduced if the plant
is harvested when it is immature or when it is overgrown,
as the foliage and stem become fibrous, brittle, pithy and
unpalatable (71). The number of days to harvest partially
depends on the harvest method. Multiple harvests are
reported to produce higher yields (5,63). Regrowth provides
up to four harvests. In Dahomey, Deutsch (63) has found
that height at first cutting affects yield. It is suggested
the the plants be cut ,at 20 cm. and that the harvest interval
should be three weeks.

Grubben (5) has reviewed and summarized harvest
techniques in various parts of the world. Harvest is
generally three to five weeks after planting. Harvests
at a later date are less desirable because of the presence
of inflorescences. Deutsch (63) has found harvest at
26 days to be the best in terms of high leaf/stem ratio.
Kamalanathan, et al., (71) note that harvest in the sixth
week after sowing gives the highest yield of edible matter.

Grain Amaranth

Studies documenting the history of grain amaranth (8,72,
73,74,75) shed little light on the crop cultural practices
used. Direct seeding takes place in the tropics at the
onset of the rainy season. Overseeding followed by thinning
as well as planting to a stand are both found in traditional
culture. In northern India and Nepal the crop is transplanted along with finger millet (*Eleusine indica*) in an inter-
crop mixture. Up to two thousand plants per hectare are

Figure 2.   A grain amaranth plant selected for non-branching and compact, clustered inflorescences.
                                        --Rodale Press Photo

randomly mixed with the millet. Both crops are hand-harvested at about 130 days from transplanting. In these areas amaranth is never grown as a sole crop.

Amaranth is a difficult crop to seed. The grain is small (.05 to .9 mg. each) (76) and must be planted shallowly to assure germination. Since the seed is placed so close to the surface, rains or irrigation can wash out recently planted fields. Transplanting has been attempted to eliminate this problem and also to assure a given planting density of strong seedlings.

A study by Mohideen and Rajagopal (66) investigated the effects of transplanting versus direct seeding on yields of *A. leucocarpus* (a synonym of *A. hypochondriacus*). Yield of transplanted crops was significantly lower than that of a direct-seeded crop. The decrease in yield was correlated with a decrease in branching of the plant and the spike.

Transplanting is a convenient and accurate method for scientific research plots, but it is not a method feasible for commercial grain production. Additional research is needed to establish adequate methods of establishing uniform grain stands without the need for transplanting or hand-seeding. Overseeding and thinning may be used but it is a less-desirable alternative.

Once the stand is established, maintenance is relatively easy. Broad leaves and erect plant habit rapidly create a closed canopy, making understory weed growth a minor problem under most conditions.

Grain amaranths can be mechanically cultivated until the canopy closes. Preliminary density trials indicate that 20,000 plants/hectare is an acceptable density for yield as well as stand management for many *A. hypochondriacus* plant types.

Information on fertilization techniques is also limited. Initial studies (28) indicate that amaranths are more responsive to fertilizer than are other cereals. Grain yields of amaranth were compared to some standard cereals. Best yields on the cereal and on amaranth were reported with a nitrogen fertilizer rate of 80 kg/ha. Wheat yields (the best of all cereals listed) were 3.61 MT/ha. compared to amaranth yields of 5.54 MT/ha. (29).

Observations from trials at the Organic Gardening and Farming Research Center indicate that maturity for most grain types is from 4 to 5 months in Pennsylvania. Presently

Figure 3.   Short-statured varietal selections of grain amaranth.   -- Rodale Press Photo

available varieties are not sufficiently uniform to be machine harvested. The central flower head may mature before the
numerous inflorescences on side branches. Where grain amaranth is used commercially it is hand-harvested, and sun dried.
Threshing and winnowing are done by hand. Ongoing breeding
work in Pennsylvania has very quickly altered plant type to
that more nearly suited to commercial handling.

Harvesting presents additional problems. The seeds
separate easily from other floral parts upon threshing, but
the small seed size makes cleaning difficult. Air separation
is essential.

Insect and disease problems are not well documented.
Holcomb (77) named a new variety of fungus *Alternaria
alternantherae* on alligatorweed, a weedy amaranth. It
has also been found to be pathogenic on several ornamental
amaranth species including *A. caudatus* (77). In Pennsylvania
symptoms similar to those described by Holcomb were observed
on several of the hundreds of accessions lines. Further
research will be needed to document these disease problems.

Insect pests have not been researched on grain species.
Early (75) mentions plant pests but there was no identification.

Crop Improvement

Amaranth offers more genetic diversity in its present
undeveloped state than do many widely-grown crops. The
wide geographic spread of the genus has resulted in the
evolution of many land races in widely separated areas.
The huge gene pool will be very important to the future
development of the crop.

Vegetable amaranth has been more thoroughly investigated
as a crop than have the grain amaranths. Selections have
been made by Asian growers for many years. Named varieties
are available from seed companies in Hong Kong, Taiwan,
and the United States which are suitable for widespread
culture.

Further domestication of grain amaranth will be a more
difficult task. Grain culture has been limited almost
entirely to subsistence farming. Grain types have been
less subjected to varietal selection as have vegetable types
(78). The major problems in grain amaranth culture are
related to seed harvest. The present harvest method, as
described earlier, is very labor-intensive. To take grain
amaranth from its semi-domesticated state, selection programs

must concentrate on several inter-related morphological characteristics, each of which will offer an improvement to more than one problem area:

1. Shorter, more stocky plant stature.
2. Resistance to lodging.
    Lodging resistance is dependent on a good branching pattern and sturdy plant frame. A desirable plant type would have upright branches which neither extend higher than the main seed head nor break when loaded with seed. Perhaps the optimal plant type will have a monocapitate morphology.
3. Higher ratio of grain to total plant biomass
    The tall plants which are most common, produce too many leaves in relation to grain and are probably not as efficient as convential grain crops. More optimal grain stover ratio types are available in germplasm collections.
4. Uniform ripening and resistance to shattering.
    The earlier the plant matures, the greater the ease of harvest. In order to be harvested mechanically, the seed must mature at the same time that the plant vegetative parts are undergoing senescence. The dehiscent utricle on the flower head, although making it easy to separate chaff from seed, causes shattering problems. Uniform maturity will reduce seed losses.
5. Large seed size.

Amaranth has good potential as a grain crop if the present problems with cultural techniques can be overcome. It is one of the few dicotyledons with this potential. This could lead to new opportunities for crop rotation, double or multicropping, or breaking up large expanses of cereals to slow spread of disease (79).

The plasticity and adaptability of the crop could lead to cropping systems in previously unusable land, particularly in semiarid regions.

Current research emphasis on amaranth is concentrated in the United States, with a few specialized projects underway in other countries. Work with vegetable amaranth consists of selecting from among available types for horticultural performance and nutritional quality. Such work is located in the Vegetable Crops Department of Cornell University, the Organic Gardening and Farming Research Center in Pennsylvania, the Federal Research Station at Mayaguez, Puerto Rico and Tamil Nadu Agriculture University in India. The Organic Center is focusing on nutritional quality of

Figure 4.   Portions of a worldwide germplasm collection being
grown at Rodale's Organic Gardening and Farming
Research Center.   --Rodale Press Photo

vegetable amaranth selections.

The greatest surge in interest in amaranth is with the grain types. The Organic Gardening and Farming Research Center is leading in this work, sponsoring collection efforts in the regions of the world where amaranth is grown. The germplasm collection is evaluated and maintained in Maxatawny, Pa. An intensive breeding program at the Center has identified higher yielding selections of a more desirable plant type with a single round of selection. Large numbers of crosses are expected to recombine many of the desirable traits from germplasm accessions. Genetic studies at the University of California at Davis and at Iowa State University promise to greatly expand the knowledge of genetic diversity within the genus.

Most research efforts are small and specialized in a limited research area. An initial symposium held by Rodale Press in 1977 brought together many of the amaranth research workers for an initial sharing of available knowledge. A comprehensive literature summary will soon be published by Rodale Press. A second symposium, planned in Pennsylvania for September of 1979, will bring together many of the world's amaranth researchers for an update of research progress.

Amaranth research has gained considerable momentum in the past three to four years. While many problems remain for the crop, including small seed size, undesirable plant types and nutritional selection needs, no serious obstacles to future development are apparent. Consumer interest in the crop is high in the U.S. among home gardeners. Seed companies report a brisk sale of vegetable and grain types. In response to a 1977 magazine request for reader cooperators in an amaranth production research effort, 15,000 home gardeners requested the research kit from Rodale Press. Over 3000 of those completed the requested crop research and observations.

Amaranth has a long way to go to make a significant contribution to the world food base, but interest in the crop is high and progress is rapid.

References and Notes

1. *Underexploited Tropical Plants with Promising Economic Value* (National Academy of Sciences, Washington, D. C. 1975).
2. G. A. Agogino, *Science Newsletter*, Washington 72, 345 (1957).

3. R. S. MacNeish, *Archaeology* <u>24</u>, 307 (1971).
4. J. D. Sauer, *Annals of the Missouri Botanical Garden* <u>37</u>,105 (1950).
5. G. J. H. Grubben, *The Cultivation of Amaranth as a Tropical Leaf Vegetable* (Royal Tropical Institute, Amsterdam, 1976).
6. J. D. Sauer, *Southwestern Journal of Anthropology* <u>6</u>, 412 (1950).
7. C. C. Black, *Weed Science* <u>17</u>, 338 (1969).
8. J. D. Sauer, *Annals of the Missouri Botanical Garden* <u>54</u>, 103 (1967).
9. P. D. Walton, *Journal of Heredity* <u>59</u>, 17 (1968).
10. M. Pal, *Proceedings of the Indian National Science Academy* <u>38</u>, 28 (1972).
11. J. D. Sauer, *Evolution* <u>11</u>, 11 (1957).
12. M. Pal, *Genetica* <u>43</u>, 106 (1972).
13. J. M. Tucker and J. D. Sauer, *Madroño* <u>14</u>, 252 (1958).
14. W. F. Grant, *Canadian Journal of Genetics and Cytology* <u>1</u>, 313 (1959).
15. M. Pal, *Proceedings of the Indian Academy of Science* <u>LX</u>, 347 (1964).
16. T. N. Khoshoo and M. Pal, *Chromosomes Today* <u>3</u>, 259 (1972).
17. M. Pal and T. N. Khoshoo, *Theor. of Applied Genetics* <u>43</u>, 252 (1973).
18. M. Pal and T. N. Khoshoo, *Theor. of Applied Genetics* <u>43</u>, 343 (1973).
19. W. F. Grant, *Canadian Journal of Botany* <u>37</u>, 1063 (1959).
20. M. J. Murry, *Genetics* <u>25</u>, 409 (1940).
21. M. Pal and T. N. Khoshoo, *Genetica* <u>43</u>, 119 (1972).
22. M. Pal and T. N. Khoshoo, *Journal of Heredity* <u>63</u>, 78 (1972).
23. J. D. Sauer, *Madroño* <u>13</u>, 5 (1955).
24. R. W. Allard, *Principles of Plant Breeding* (John Wiley and Sons, N. Y.), <u>472</u> (1960).
25. M. Pal and T. N. Khoshoo, *Z. Pflanzenzuchtg.* <u>78</u>,135 (1977).
26. P. S. Misra, M. Pal, C. R. Mitra and T. N. Khoshoo, *Proceedings of the Indian Academy of Science* <u>LXXIV</u>, 155 (1971).
27. J. Elias, *Proc. First Amaranth Seminar* (Rodale Press, Inc., Emmaus, Pa. 1977) p. 16.
28. D. Schmidt, *Proc. First Amaranth Seminar* (Rodale Press, Inc., Emmaus, Pa. 1977) p. 121.
29. M. J. Wolf, M. M. MacMasters and C. E. Rist, *Cereal Chemistry* <u>27</u>, 219 (1950).
30. M. M MacMasters, P. D. Baird, M. M. Holzapfel and C. E. Rist, *Economic Botany* <u>9</u>, 300 (1955).

31. P. V. Subba Rao and K. J. Goering, *Cereal Chemistry* <u>47</u>, 655 (1970).
32. C. R. Smith, Jr., M. C. Shekelton, I. A. Wolff and Q. Jones, *Economic Botany* <u>13</u>, 132 (1959).
33. F. W. Martin and R. Ruberte, *First Amaranth Seminar* (Rodale Press, Inc., Emmaus, Pa. 1977) p. 105.
34. J. S. Mugerwa and R. Bwabye, *Trop. Grasslands* <u>8</u>, 49 (1974).
35. S. K. Imbama, *East Afr. Agr. Forest J.* <u>39</u>, 246 (1973).
36. O. L. Oke, *West Afr. Sci. Assoc.* <u>2</u>, 42 (1966).
37. M. Fafunso and O. Bassir, *J. Food Sci.* <u>41</u>, 214 (1976).
38. R. Carlson, *Proc. First Amaranth Seminar* (Rodale Press, Inc., Emmaus, Pa. 1977) p. 83.
39. X. R. Rajkumar, K. Durairaj and C. Gnanadickam, *Current Science* <u>42</u>, 317 (1973).
40. D. R. Schmidt and W. C. Kelly, *Communications in Soil Science and Plant Analysis* <u>4</u>, 95 (1973).
41. G. S. Gilbert,H. F. Eppson, W. B. Bradley, and O. A. Beath, *Bulletin no. 277, Nitrate Accumulation in Cultivated Plants and Weeds,* (U. Wyoming Agr. Exp. Sta. Laramie, Wyoming 1946).
42. D. N. Maynard, A. V. Barker, P. L. Minotti, and N. H. Peck, *Advance Agr.* <u>28</u>, 71 (1976).
43. O. A. Lorenz, in D. R. Nielson and J. G. MacDonald, eds., *Nitrogen in the Environment V. 2* (Academic Press, New York 1978) p. 201.
44. J. W. White, Jr., *J. Agr. Food Chem.* <u>23</u>, 886 (1975).
45. S. R. Tannenbaum, D. Fett, V. R. Young, P. D. Land, and W. R. Bruce, *Sci.* <u>200</u>, 1487 (1978).
46. D. R. Nielson and J. G. MacDonald, eds., *Nitrogen in the Environment V. 2* (Academic Press, New York 1978).
47. J. M. Kingsbury, *Poisonous Plants of the United States and Canada* (MacMillian, New York, 1964).
48. A. Hodgkinson, *Oxalic Acid in Biology and Medicine* (Academic Press, New York, 1977).
49. L. Hagler and R. H. Herman, *Amer. J. Clini. Nutrition* <u>26</u>, 758, 882, 1006, 1073, 1242 (1973).
50. A. A. Hoover and M. C. Karunairatnam, *Biochem. J.* <u>39</u>, 237 (1945).
51. P. P. Singh, L. K. Kothari, D. C. Sharma and S. N. Saxena, *Nutrition* <u>25</u>, 1147 (1972).
52. P. P. Singh, *Quality Plant Materials Vegetables* <u>22</u>, 335 (1973).
53. K. P. Basu and D. Ghosh, *Indian J. Med. Res.* <u>31</u>, 37 (1943).
54. S. Premakumari, G. Geetha, and R. P. Devadas, *Indian J. Nutrition Dietetics* <u>15</u>, 67 (1978).
55. U. Pingle and B. V. Ramasastri, *Brit. J. Nutrition* <u>39</u>, 119 (1978).

56. R. M. Welch, W. A. House, and D. Van Campen, *J. Nutrition* 107, 929 (1977).
57. D. R. Schmidt, H. A. MacDonald, and F. E. Brockman, *Agron. J.* 63, 559 (1971).
58. N. Krishnamurthy, S. Geetha, and R. P. Devadas, *Indian J. Nutrition Dietetics* 13, 293 (1976).
59. R. Hegnauer, *Chemotaxonomie Der Pflanzen* Band 3 (Birkhauser Verlad, Basel and Stuttgart, 1964) p. 81.
60. W. H. Lewis and P. F. Elvin-Lewis, *Medical Botany* (John Wiley, New York, 1977) p. 33, 69, 96).
61. P. J. Smyth, *Irish Vet. J.* 31, 175 (1977).
62. P. R. Cheeke, J. H. Kinzell and M. W. Pederson, *J. Anim. Sci.* 46, 476 (1977).
63. J. Deutsh, *Genetic Variation of Yield and Nutritional Value in Several Amarantus Species Used as a Leafy Vegetable*, PhD. Thesis, Cornell U. (1971).
64. J. A. Samson, *Surinaamse, Landbrouw* 20, 15 (1972).
65. J. E. Knott and J. R. Deanon, Jr., *Vegetable Production in Southeast Asia* (Los Baños, Laguna, Phillipines, 1967).
66. M. K. Mohideen and A. Rajagopal, *South Indian Horticulture* 23, 87 (1975).
67. E. L. McWilliams R. Q. Landers, and J. P. Mahlstede *Ecology* 49, 290 (1968).
68. J. M. Baskin and C. C. Baskin, *Oecologia* 30, 377 (1977).
69. C. R. Evans, *Botanical Gazette* 73, 213 (1922).
70. I. C. Onwueme and S. A. Adegoroye, *Journal of Agricultural Science* 84, 525 (1975).
71. S. Kamalanathan, S. Sundararajan, A. Shanmugham, R. Subbiah and S. Thamburaj, *South Indian Horticulture* 18, 77 (1970).
72. J. D. Sauer, *Southwestern Journal of Anthropology* 6, 412 (1950).
73. J. D. Sauer, *Annals of the Missouri Botanical Garden* 37, 561 (1950).
74. W. E. Safford, *Proceedings 19th Intern. Congress Americanists*, p. 286, (1915).
75. D. Early, *Proc. First Amaranth Seminar* (Rodale Press, Inc., Emmaus, Pa. 1977) p. 42.
76. H. Hauptli, *Proc. First Amaranth Seminar* (Rodale Press, Inc., Emmaus, Pa. 1977) p. 71.
77. G. E. Holcomb, *Phytopathology* 68, 265 (1978).
78. H. Hauptli and S. Jain, *California Agriculture* 31, 6 (1977).
79. H. Singh, *Grain Amaranths, Buckwheat and Chenopods* (Indian Council Ag Res, Cereal Crop Series, No. 1, New Delhi, 1961).

4

The Buffalo Gourd

A Potential Arid Land Crop

W.P. Bemis, James W. Berry and Charles W. Weber

Abstract

A plant indigenous to western North America, the Buffalo gourd, has evolved characters that give this plant potential as an arid lands crop. It is capable of producing acceptable yields of vegetable oil, protein, and starch. The excessive vine growth may also have a potential as forage. An intensive research project supported by disciplines of genetics and plant breeding, biochemistry, and nutrition and toxicology are engaged in a study of domesticating this species.

Introduction

The feral xerophytic. Buffalo gourd, Cucurbita foetidissima HBK, has evolved in the semi-arid regions of western North America and is well adapted to desert environments. During evolution from its Cucurbita progenitors the Buffalo gourd has developed several unique aspects, presumably without much influence of man, which give it potential as a cultivated crop. This plant is capable of producing a high quality vegetable oil and an acceptable protein in the seeds, a high yield of starch in its extremely large storage roots, and prodigious vine growth with forage potential. Thus this wild perennial has the potential of becoming a crop adapted to arid to semi-arid lands, producing additional food critically needed to feed peoples of these regions.

***

Pre-history. The Buffalo gourd has been associated with the American Indians for upwards of 9,000 years, most likely as a camp follower rather than a truly domesticated species (1). There are many reports of Buffalo gourd seeds being used as food. They were reportedly ground fine and made into mush, or roasted in various ways. The oil from ground seeds was extracted in hot water and used as a cosmetic. The mature green fruit were, and still are, used as a detergent for cleaning and scouring. The root was used mainly as either a source of detergent or in various medicinal ways. It is very doubtful that the American Indians ever cultivated the Buffalo gourd, but rather harvested parts of the plant from wild colonies. With the exception of the seeds, all parts of the plant are extremely bitter and must be processed in some manner before they can be used as food.

Taxonomy. According to Bailey (2) Cucurbita foetidissima was first described in 1817 and was later mistakenly named Cucumis perennis by James (1820), who found it growing from "arid and sandy wastes, along the base of the Rocky Mountains, from the confluence of the Arkansas, and Boiling Spring Fork, to the sources of the Red River." Asa Gray transferred it in 1852 as Cucurbita perennis, and under this name it remained until 1881 when Cogniaux brought up the name Cucurbita foetidissima of Humbolt, Bonpland and Kunth, having seen the Bonpland specimen in Paris. It is known by many common names such as Missouri Gourd, Calabazilla Loca, Chili Coyote and Fetid Gourd. The vines often emit a pungent odor, hence the name foetidissima. However, the common name by which it is most widely known is Buffalo Gourd.

Interspecific Relationships. A numerical taxonomic study of Cucurbita species relationships (3) has shown that C. foetidissima is not phenotypically similar to any other of the 20 species or species groups that were studied. Bemis and Whitaker (4) describe four restricted xerophytic species, C. cylindrata Bailey, C. cordata Wats., C. palmata Wats., and C. digitata Gray, known collectively as the digitata group, and although a fifth species, the wide ranging and variable C. foetidissima is truly xerophytic, it is only distantly related to the species of the digitata group.

Many attempts have been made to hybridize the Buffalo gourd with other Cucurbita species, but the only reported success was using it as a pollen parent with Cucurbita moschata, a domesticated species (5,6). In both cases the underdeveloped embryos had to be cultured and the hybrid, although vegetatively vigorous was completely sterile. Bemis (7) successfully produced the amphidiploids of the hybrid and restored female fertility. The diploid hybrid at metaphase I configuration of meiosis contained from three to five loosely associated bivalents and 30 to 34 univalents (n=20). The amphidiploid contained 40 bivalents at metaphase I. However, the particular C. moschata cultivar used in the initial cross contained a genetic factor which, when combined with the C. foetidissima genome, resulted in genetic male sterility. Other C. moschata cultivars which do not contain this incompatibility factor have been found and another attempt at producing fertile amphidiploids has been initiated.

Plant Morphology. The Buffalo gourd is perennial by virtue of its exceedingly large fleshy storage roots. A single root may reach fresh weights of over 40 kg. in three or four seasons of growth. The frost-sensitive vines are killed by below $0^\circ C$ temperatures, but the roots may survive winter air temperatures as low as $-25^\circ C$, particularly when the soil has the insulation of snow cover. The plants' primary mode of reproduction is asexual by the development of adventitious roots produced at the nodes of the vines. Large homogeneous colonies of plants are produced in this manner. The annual vine growth is extensive. A single plant in New Mexico, age unknown, produced a total linear vine length of 220 m in a five month growing season (8).

The large unisexual flowers (pistillate or staminate) are borne singly at most nodes of the vine. The predominate sex expression of the plant is monoecious, pistillate and staminate flowers being produced on the same plant. However, a dominant mutant gene causing abortion of the male flower buds resulting in gynoecious (all female) plants was found to be widespread in native populations and will be discussed in detail under Seed Production. The fruit (pepos) are usually round with diameters of 5-7 cm. The number of seed per fruit ranges from 200-300, with an average weight of 4 g per 100 seed. A single plant is capable of producing up to 200 fruit in a single growing season. The seed contains 30-40 percent edible oil and 30-35 percent protein.

The vine growth is extremely prolific. A single root in its second season of growth will produce from 6 to 20 vine initials which are capable of producing vines of 6 m or more in length. These initial vines also branch at many of the nodes resulting in a dense mat of vine growth. The vines are ground cover vines, not climbing vines. The harsh, sandpaper-like leaves are usually entire, ovate to sagittate, with a base width of 10-13 cm and a mid-rib length of 20-25 cm. The single native plant studied by Dittmer & Talley (8) having a total vine length of 220 m was calculated to have about 15,000 leaves. The photosynthetic capability of these plants must be tremendous.

Research History. Dr. Lawrence C. Curtis, a recognized authority and outspoken advocate of Buffalo gourd as a crop, first described the potential of this native species in 1946 (9). Based on native stands of this species he made 4 points: (a) the plants are perennial; (b) they grow on wastelands in regions of low rainfall; (c) they can produce an abundant crop of fruit which contains seed rich in oil and protein; and (d) the fruit lends itself to mechanical harvesting. Curtis concluded his report with the statement that it seems ironic that the answer to some of the problems of our under-nourished populations may be growing in wide areas in their immediate vicinity as a neglected weed.

Shahani et al. (10) made the first detailed chemical study of the oil extracted from seeds of the Buffalo gourd. The fatty acid composition of the oil was found to be acceptable, the major components being linoleic 65.3%, oleic 23.0%, palmitic 6.13% and stearic 2.22%. They reported that the crude oil was dark in color and exceedingly resistant to bleaching. This characteristic was attributed to change in the seed pigments due to weathering of the fruit in the field. Refining, bleaching, and deodorization, however, gave a bland oil with good stability and no tendency to revert to poor flavor. Field plantings (in Texas) were limited (6 plots, each of 1/500 acre) but it was concluded that there was such a wide variation in number of seeds per fruit and number of fruits per plant, that there would be unlimited opportunity to improve yield through selection, presuming that these traits are heritable.

Paur (11) studied some small plots of the Buffalo gourd in New Mexico. The plants grew well on a light mesa soil, had low water requirements, and were free from diseases and insect injury.

An extensive report (12) was published in 1974 covering 6 years of field research to domesticate the Buffalo gourd as part of a project sponsored by the Ford Foundation in Tel Amara, Lebanon. (The program was terminated with the retirement of Dr. Curtis, and the subsequent civil war in Lebanon.) The initial germplasm source upon which the work in Lebanon depended came from a restricted genetic base from a single collection from a site in Texas. Despite this narrow genetic base, the results indicate a vast store of genetic variation within this species. For example, fruit yield from 712 single 2-year old plants, spaced 3 x 3 m, was as follows: 60 plants produced 0 fruit; 533 plants produced from 1-100 fruit; 104 produced from 101-200 fruit, and 15 produced from 201-300 fruit. The highest yielding plant produced 271 fruit. The crude fat from seed of 50 selected plants ranged from 25.6% to 42.8% and the protein from 25.9% to 35.0%. Many plants with unusual characters ranging from vine habit to sex expression were observed, again indicating the enormous amount of genetic variation found in this species.

Studies (13,14) on the nutritional value of the seed

protein indicated that it compares favorably with protein from other oil seed crops.

University of Arizona Program

Cucurbita foetidissima has been part of a research project at the University of Arizona since 1963, but only as one of several species of Cucurbita used to study the genetic barriers to interspecific hybridization. In 1973, at the urging of Dr. L.C. Curtis and others, a movement was commenced at the University of Arizona to explore the possibility of initiating a full scale research project on domestication and utilization of the Buffalo gourd. A committee was appointed in the College of Agriculture to study the feasibility of this investigation and it recommended the initiation of such a study. With the assistance of the Agency for International Development, a technical series bulletin was published (15).

The project was conceived as covering a broad spectrum including breeding, domestication, and utilization. Obviously, such a broad based program called for an interdisciplinary approach to these problems. The initial major disciplines involved were 1) genetics and plant breeding, 2) biochemistry and 3) nutrition and toxicology (16).

In order to study the agronomics of the Buffalo gourd as a new field crop, the first major objective was the creation of a relatively homogeneous seed source from which cultural plots would be established. The seed originally collected from wild colonies was extremely heterogeneous. In addition to the need for relatively homogeneous test seed, it should be representative of the best available germplasm for crop production.

Plant Collections. The current range of the feral Buffalo gourd extends over nearly 3 million square kilometers in North America. The north-south axis of its range extends from Guanajuato in central Mexico to near the southern South Dakota border in the United States, a distance of about 2,700 kilometers. Presumably the major stands of plants extend from the Chihuahuan northward along the eastern base of the Rocky Mountains. A westward branch extends through New Mexico into south-central Arizona. The eastern extension of its range almost to the Mississippi River is probably recent since the plants are found mainly along highways or railroad right-of-ways. Its most westward advance into the coastal mountains of southern California is also probably recent. It can be considered both a ruderal (or roadside) colonizer, and an agrestal (or agricultural weed).

The mature gourds are attractive and were most likely carried by persons along their travels which greatly extended the current native range of the Buffalo gourd.

Regardless of how its current range was developed, it represents the genetic variation that exists in this species. An effort was made to bring a cross-section of this genetic diversity into one location through plant collection trips and the establishment of germplasm nurseries.

Table 1 shows the number of accessions collected from three trips made in 1975, 1976 and 1977 and the three germplasm nurseries (GPN) established from the collected accessions.

Table 1.   Accessions used in seeding the 3 germplasm nurseries (GPN) tabulated by collection site.

Collection site	No. of accessions			
	GPN-76	GPN-77	GPN-78	Total
Arizona	43	1	3	47
New Mexico	2	17	4	23
Texas	6	10	5	21
California	–	1	10	11
Mexico	–	10	–	10
Nebraska	–	9	–	9
Kansas	–	6	–	6
Oklahoma	–	6	–	6
Colorado	–	6	–	6
Utah	3	–	2	5
Illinois[1]	–	1	–	1
TOTAL	54	67	24	145

[1]Five adventitious roots sent from Illinois have been established in GPN-77.

Some of the accessions shown in Table 1 were not from the collection trips per se, but were sent in by interested cooperators.

Germplasm Nurseries. The purpose of the germplasm nurseries (GPN) is to establish a population of Buffalo gourd plants which would represent a cross-section of the genetic diversity found in the range of its native habitat.

The degree of variation found in some of the original collections is shown in Table 2 (17). The data in Table 2

are from fruit representing 85 accessions or collection
sites, 29 in Arizona, 19 in New Mexico, 12 in Texas, 7 in
Nebraska and 6 each in Colorado, Kansas and Oklahoma.

Table 2. Means, ranges and coefficients of variations of
seven fruit and seed characters of 85 Buffalo
gourd accessions.

Characteristic	Mean	Range	CV
Fruit diameter	6.5 cm	5.2 - 7.7	9.1
Seed wt./100 seeds	3.8 g	1.1 - 5.5	20.0
Seet wt./fruit	8.4 g	2.7 - 18.8	36.9
Seed number/fruit	225	87 - 386	28.3
% Embryo in seed	67.3	42.1 - 76.3	9.7
% Crude fat in seed	32.9	21.1 - 43.1	14.3
% Crude protein in seed	30.7	19.5 - 35.4	8.7

The GPNs were established at the Agricultural Experiment
Station of the University of Arizona, Tucson, Az. Seed for
each accession was hand planted in 15 hills 1 m apart on rows
2 m apart and subsequently thinned to 1 plant per hill. A
total of 145 accessions were seeded in this manner. These
GPN now serve as a source of material to initiate breeding
programs.

Seed Production. The Buffalo gourd has been classified by
taxonomists as being monoecious, which is consistant with all
of the known species of Cucurbita. Observations of plants in
GPN-76 during their second growing season indicated that many
of the accessions were segregating two types of sex expression. Forty-seven accessions, having a population of 10 or
more plants, were classified as having all monoecious plants
(pistillate and staminate flowers) or were segregating monoecious and gynoecious plants (pistillate flowers only).
Twenty-five accessions were entirely monoecious, however 22
accessions were segregating monoecious and gynoecious plants
and were randomly scattered throughout the collections. The
total count for sex type of the 22 segregating accessions was
153 monoecious to 118 gynoecious plants. A tentative hypothesis suggests that the gynoecious expression is conditioned
by a dominant gene restricted to the heterozygous state and
the monoecious expression by the homozygous recessive state,
i.e. Aa = gynoecious sex and aa = monoecious sex. A suggested term for this type of sex expression is "monogyno-
dioecy" indicating populations which contain monoecious and
gynoecious plants in approximately equal numbers.

Curtis (12) observed a similar, or possibly the same, type of sex expression in his Buffalo gourd populations in Lebanon. He referred to this mutant type as "antherless" and noted it segregated in a 1:1 ratio when crossed with a monoecious plant. He also reported that the mean fruit yield per plant for his 10 best "antherless" plants was 272 fruit, while the mean fruit yield per plant of his 6 best monoecious plants was only 209 fruit.

Since a major objective of the domestication program was the creation of a relatively homogeneous seed supply, it was decided to rapidly produce "hybrid seed". After one generation of selection, the two most promising lines derived from gynoecious plants were selected based on fruit and seed yield and crude fat percentage in the seeds. Seed of these lines coded as 158-2 and 142-1 were selected from gynoecious plants and were expected to segregate 1:1 for monoecious and gynoecious plants. It was decided to produce the "hybrid seed" in a reciprocal fashion, i.e. two isolated seed production plots would be established. The success of these seed production plots was dependent on both selections segregating 1:1, monoecious: gynoecious. In one isolation seed production plot the female or seed parent would be rogued so that only gynoecious plants would be left. The male or pollen parent would be rogued so that only monoecious plants would be left. The male flowers would be the source of pollen to fertilize the flowers on the gynoecious line. The seed then extracted from fruit on the gynoecious plants would be of hybrid origin.

Both selections segregated about 1 monoecious to 1 gynoecious and the 2 isolation seed production plots were established by roguing the monoecious segregates from the seed parent and gynoecious segregates from the pollinator parent. The segregation ratios were as follows: 158-2 segregated 167 monoecious and 126 gynoecious with 58 plants unclassified. Selection 142-1 segregated 61 monoecious and 65 gynoecious with 13 plants unclassified. The unclassified plants were rogued before they flowered and most were probably gynoecious which normally are delayed in flowering compared to monoecious plants.

The major isolated seed production plot (152-2 x 142-1) contained 114 gynoecious plants which during the first season yielded about 3,500 fruit and 23 kg of cleaned seed. This seed is now available for distribution to interested persons, as well as being used to establish the preliminary cultural plots for the research project. Even though the yield from the seed production plots were greater than anticipated in the first season of production, it was observed that early in the second season (1978) the first fruits on the seed parent line failed to set. The pollinators were absent. In the native range of the Buffalo gourd, the primary pollinators are the solitary squash and gourd bees, **Peponapis** and **Xenoglosa** (18). Hives of honeybees were brought into the area of the isolated seed production plots, but they foraged on more attractive species. These observations point out a potential problem in large area production fields of Buffalo gourds where a large population of pollinators must be present for adequate fruit set and production of seed.

Cropping System for Buffalo Gourds. Three factors make the Buffalo gourd unique as a crop plant. These are, 1) its perennial habit, 2) its asexual mode of reproduction, and 3) its yield components, fruit (seed) for oil and protein, its root for starch, and possibly the vine as a forage for domestic animals. No other crop to our knowledge has these three attributes.

It is proposed that a Buffalo gourd field be established by direct seeding (Figure 1) using hybrid seed with an initial "in row" spacing of 15 to 30 cm with rows 1-2 m apart. This plan would give an initial plant stand varying from about 6000 to 1500 plants/ha. At the end of the first growing season a destructive vine harvest would be made using some type of vine seed thresher to recover the seeds from the fruit (Figure 2). Depending on the extent of vine growth, the remnant vines may or may not be left in the field, or could be grazed by sheep or cattle. The crowns of the perennial roots will be 7-10 cm below the soil surface and would be protected.

The Buffalo gourd is frost sensitive and the vines of mature plants are usually destroyed when minimum temperatures approach 4°. Once the vines have been destroyed by low temperatures, the roots remain dormant as long as night temperatures remain 5 to 7° or less. Under Tucson, Arizona conditions, new growth of the vines is initiated in late February. Flowering and fruiting begins around late March and continues through October. The fruit remains in a harvestable condition for several months after the vines have been destroyed.

At the second growing season, the plant stand will be increased by a factor of 5 to 10 depending on the amount of vine growth and adventitious rooting (Figure 3). At the end of the second season there will be another destructive vine harvest to recover the seed. Then sometime during the dormant season, alternate 1 m swaths of the field will be mechanically dug for roots. This will accomplish 2 purposes; the harvesting of roots for starch; and the thinning of plants to keep them from over-crowding.

Fig. 1.   Buffalo gourd seedling 6 weeks from seeding.

Fig. 2.   Fruit set on Buffalo gourd vines.

Fig. 3.   A field of two-year-old Buffalo gourd vines.

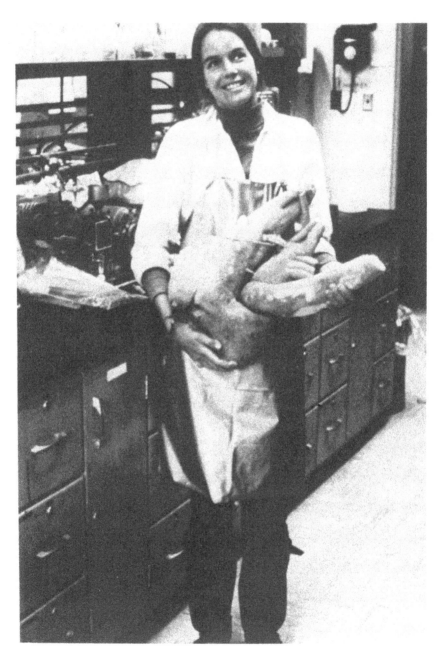

Fig. 4.   A 6 kg. two-year-old Buffalo gourd root being held
by laboratory technician Susan Kunz.

Roots of two year old plants were monitored for changes in starch content (19) on 17 dates throughout the 1976 calendar year. In general, the starch content of the root was relatively constant except for April 14 when it dropped to slightly over 20% (fresh weight basis) and hit a low of 18.5% on May 3. These dates coincide with the very active vine growth and initial fruit development. The starch content was quickly replenished reaching over 40% on June 1 and peaking at 52% on August 15. For acceptable starch content, it is reasonable to harvest the roots any time during the late growing or dormant season, after the fruit harvest (Figure 4).

The third season the alternate 1 m swaths would be dug. In the meantime, the original swath would be regenerated by asexual rooting of the vines. This procedure would continue as long as the vines remained productive. Thus, the crop would yield oil and protein from seed harvest, starch through root harvest, and possible forage by feeding the remaining vines.

To date, no reliable yield data have been obtained. A relatively homogeneous seed source has been developed, Arizona Hybrid No. 1, and this seed is being used to establish cultural plots. Since the Buffalo gourd is a perennial crop, several seasons of testing will be necessary to obtain valid yield data.

If the Buffalo gourd will produce 30 fruit per m^2 and each fruit produces 10 g of seed, then a yield of 3000 kg/ha per year of seed can be anticipated. Observations on single plants seem to indicate that this projected yield can be realized. Assuming each seed is composed of 35% oil, the yield of oil would be 1050 kg/ha. Yield of protein is estimated at 450 kg/ha, assuming a content of 15% recoverable protein.

Root yield can be estimated by assuming that a square meter of soil surface, mechanically dug to a depth of 30 cm will yield 9 kg of roots. Assuming 15% starch in these roots, the yield of starch per hectare is 13,500 kg.

Current Status of Field Plot Research. The field plot research on the Buffalo gourd is centered at the Arizona Agricultural Experiment Station in Tucson, and Mesa, Arizona. At the Tucson station there are 3 germplasm nurseries established, a breeding plot designed primarily to study the inheritance of sex-expression, three isolated seed production plots which will produce seed of seven experimental hybrids, and a 1/3 hectare preliminary cultural plot seeded to Arizona Hybrid No. 1.

A field plot seeded in 1977 at the Mesa Station contains 32 selections of superior plants from the germplasm nursery. A plot has also been seeded to Arizona Hybrid No. 1 at the Mesa Station to study the consumptive use of water by the Buffalo gourd.

It is planned to establish a 5 hectare plot in 1979 primarily for cultural studies seeded mainly to Arizona Hybrid No. 1, but will also include trial plantings of the six other experimental hybrids as well as the three monoecious pollen parent lines.

Nutritional and Biochemical Research. The seeds and roots of this potential crop hold the key to its ultimate value. Utilization will depend on an understanding of the nature of the oil and protein in the seed and starch in the roots.

Seed Studies. Seeds account for one-third of the pepo weight and are composed of one-third seed coat and two-third embryo. The proximate analysis of the seeds and seed components is shown in Table 3. The seeds would have value as a component of ruminant feeds.

Table 3.   Proximate analysis of Buffalo gourd seed.

Material Analyzed	Moisture %	Crude Protein %	Crude Fat %	Acid Detergent Fiber %	Ash %
Whole seed	4.9	32.9	33.0	35.1	3.1
Seed coat	7.0	20.5	1.3	82.5	1.0
Embryo	4.5	37.5	48.0	6.9	4.2
Embryo (defatted)	4.1	75.0	1.0	11.8	7.6

They could be incorporated in limited amounts into the diets of monogastric animals such as poultry and swine, allowing direct and inexpensive utilization. The gross energy value of 6.17 kcal/g is relatively high. Assuming a loss during digestion of 30%, the available metabolizable energy would be 4.3 kcal/g, making Buffalo gourd seed comparable to other oil seeds in this respect (20).

Seeds and seed components have been evaluated for protein quality by animal feeding trials (21). Weanling CD-1 mice with an average body weight of 9 g were employed in the experiment. They were housed in suspended stainless steel cages and were given feed and water ad libitum during a three week period.

The isocaloric isonitrogenous diets were in powder form, containing 8% protein and providing 3.70 kcal/g. Whole egg was used as the standard. Protein quality was assessed by determining protein score, protein efficiency ratio (PER), net protein retention (NPR), and essential amino acid index (EAAI). A protein score of 28 and an EAAI of 62 were found for whole Buffalo gourd seed. The first limiting amino acid was calculated to be methionine. The amino acid patterns of Buffalo gourd and whole egg are shown in Table 4, and values for PER and NPR are given in Table 5. An NPR value greater than 2.00 indicates reasonably good protein quality. When seed coats, which contain about 20% crude protein, were fed as the sole source of protein; a net loss of weight, a negative PER and an NPR below 1.00 resulted.

Table 4.   Essential amino acid composition as percent of protein.

Amino Acid	Whole Egg	C. foetidissima
Lysine	6.3	4.5
Histidine	2.2	2.2
Arginine	6.7	13.0
Threonine	4.8	2.0
Valine	7.6	3.8
Methionine + cystine	7.6	1.7
Isoleucine	8.7	3.3
Leucine	9.5	5.7
Phenylalanine + Tyrosine	10.4	7.4

Table 5.   Results of feeding experiments with weanling CD-1 mice measuring protein quality of Buffalo gourd seed and its components.

Material Fed	Gain in Body Weight (g)	PER	NPR
Whole egg	13.6	2.50	4.00
Buffalo gourd			
Whole seed	3.1	1.50	2.20
Embryo (defatted)	8.8	1.74	3.20
Hulls	-2.2	-0.93	0.86
Oil	11.1	-	-

This indicates that the protein in that seed fraction is not available to monogastric animals.

Analysis of the seed coats of Buffalo gourd and other oilseeds is shown in Table 6.

Table 6.   Seed coat composition of Buffalo gourd and other oilseeds.

Species	Acid detergent fiber %	Acid detergent Lignin %	Crude protein %	Hemi cellulose %	Cellulose %	Lignin %
Cucurbita foetidissima (Buffalo gourd)	61.5	25.8	17.3	10.0	26.0	29.2
Gossypium hirsitum (cotton)	64.5	21.8	5.4	21.0	35.0	21.8
Helianthus annus (sunflower)	61.1	27.8	3.9	16.0	39.0	27.8

Calculations based on the equation of Goering and Van Soest (22) predict suitable digestibility of Buffalo gourd seed coats when utilized as a feed component for ruminants. In comparison, cottonseed hulls are a valuable ingredient of ruminant feeds although the protein level is much lower.

In a study of cucurbit seed coat hemicelluloses (23), Buffalo gourd seed coat material was subjected to a series of extractions and a delignification step. It was shown that virtually all of the protein was associated with two hemicellulose fractions. One fraction which contains 13% protein was dissolved by 10% base, while the other was dissolved by the same solvent only after delignification, which destroys the protein present. These observations suggest that only ruminant microflora can make Buffalo gourd seed coat protein available for digestion.

The effect of amino acid supplementation on the protein quality of seed and seed fractions in mice has been investigated recently (24) and the data are shown in Table 7. The body weight gains for Buffalo gourd seed supplemented with lysine, methionine, threonine, and valine were equivalent to those obtained with the whole egg control. (Single amino acid supplementations with threonine produced the second largest weight gain). Threonine and methionine were shown to be the first and second limiting amino acids, respectively. The PER for the diet supplemented with all four

Table 7. Body weight gains, PERs and NPRs for Buffalo gourd seed supplemented with amino acids.

Treatment	Body weight gain (g/day)	PER (corrected)	NPR
Whole egg	0.44c[1]	2.50c	4.02c
Buffalo gourd whole seed	0.22a	1.21a	2.31a
B.g. + methionine	0.24ab	1.24a	2.56ab
B.g. + lysine	0.21a	1.17a	2.32a
B.g. + valine	0.21a	1.07a	2.28ab
B.g. + threonine	0.31b	1.45a	2.48ab
B.g. + meth + lys	0.25ab	1.28a	2.40ab
B.g. + meth + thre	0.27ab	1.35a	2.72ab
B.g. + meth + thre + lys + val	0.46c	2.04b	3.00b

[1] Values with different letters are significantly different at the 0.05% level.

amino acids was lower than the control, suggesting that other amino acids (isoleucine and tryptophan) may be required to achieve equality with whole egg control.

Small batches of seed (five kilograms) have been successfully processed to yield oil, defatted flour, and a seed coat fraction (25). Seed were decorticated in a Bauer mill, flaked, extracted with pentane, ground and screened. The defatted flour had a protein content of 70%, and it was used to prepare a protein concentrate by dissolution in base and precipitation by acid in the isoelectric region. Seed meals, flours, and protein concentrates such as these range from 50-85% in protein content and may prove to be valuable sources of supplemental protein for feeds and foods.

Defatted embryo material similar in composition to the defatted flour just described has been evaluated as a protein source in feeding experiments with mice (24). Body weight gains on the defatted embryo diet were close to those for soybean meal. Both were below those obtained with diets based on soybean meal supplemented with methionine. Defatted embryo had virtually the same PER as soybean meal and cottonseed meal, which are both extensively used as protein sources.

The presence of proteolytic enzyme inhibitors, hemagglutinins, and other nutritional antagonists in legumes has been studied extensively. Such substances occur commonly and con-

tribute significantly to poor digestibility and other physiological effects caused by inadequate cooking. Research on
soybeans (26) has contributed much of the knowledge on these
anti-nutritional factors. The possibility of C. foetidissima
seeds becoming a protein source for humans and monogastric
animals requires that appropriate investigation as to the
presence of these factors be carried out. Feeding experiments in which seed and seed fractions have been the sole
source of protein have not shown evidence of the presence of
proteinase inhibitors or hemagglutinins. In vitro experiments comparing Buffalo gourd seed and soybeans have demonstrated the presence of trypsin inhibitors and hemagglutinins at a level approximately one-tenth that of soybean. The
presence of phytate in Buffalo gourd seed (27) has been reported along with a way to separate it in the preparation of
defatted flour. Using an improved analytical method for phytate (29) we have found 1.51% in the defatted embryo flour.

The oil in the seed can be isolated readily by solvent
extraction or by mechanical pressing. Values for oil content
in seeds, as analyzed by several investigators, have generally been near 30%. Curtis (12) found large variation among
50 different individual plant sources. The level of oil in
decorticated cucurbit seeds was calculated to be 49.5 $\pm$ 2.3%
(29). Eighty-five accessions of C. foetidissima were assayed
for phenotypic variation (17) and the percent crude fat had a
mean of 32.9%, with a range of 21.1-43.1% (Table 2). The
means, ranges, standard deviations, and coefficients of variability for fatty acids were also determined, and sufficient
variation in oil composition was found to suggest possible
alteration through plant breeding. Oleic and linoleic acids
displayed strong negative correlation.

The fatty acid distributions in selected edible oils
are presented in Table 8. There is close similarity between
corn, soybean, and Buffalo gourd oils, suggesting the potential of the latter for food purposes. Certain cucurbit
oils have been used for cooking purposes in other countries
(9,30).

Unrefined Buffalo gourd oil has been evaluated in a
feeding study with weanling mice (21). The oil was incorporated in isocaloric isonitrogenous diets in amounts ranging up
to 11 percent of the total diet (see Table 5). Even at the
highest level, normal growth occurred with no evidence of deleterious effects.

Linoleic acid is the predominant fatty acid in the oil
(17), ranging from 50-69%. This assumes a very favorable
S:M:P ratio and enhances the oils' nutritional appeal.

Table 8. Fatty acids of various vegetable oils.

Oil	14:0[1] Myris- tic	16:0 Palmitic	18:0 Stearic	18:1 Oleic	18:2 Lin- oelic	18:3 Lin- oelic	S:M:P[2] Ratio
Buffalo gourd	T[3]	9.7	4.5	27.0	58.5	T	1:1.9:4.1
Corn	T	13.0	4.0	29.0	54.0	T	1:1.7:3.2
Cotton	1.0	29.0	4.0	24.0	40.0	T	1:0.7:1.2
Peanut	T	6.0	5.0	61.0	22.0	T	1:5.5:2.0
Saf- flower	T	8.0	3.0	13.0	75.0	1.0	1:1.2:6.8
Soybean	T	11.0	4.0	25.0	51.0	9.0	1:1.7:3.4

[1]Ratio of total carbon atoms to number of carbon-carbon double bonds

[2]Ratio of saturated to monounsaturated to polyunsaturated fatty acids

[3]Less than 1%

Current investigation in our laboratory indicates that the crude oil can be purified by current refining methods. Degumming, caustic refining, bleaching, and deodorization steps are being studied to select conditions which will produce the best edible oil in terms of quality and process efficiency.

Root Studies. Starch is the major component in the roots, and it is present in a potentially commercial amount. Domestication would provide a source of starch from a xerophytic plant. Current sources of starch are tropical or mesophytic plants.

The many food and industrial uses for starch make it a valuable item of commerce. It can be hydrolyzed by enzymatic or chemical methods to dextrins, maltose, and glucose, all of which are utilized in foods, beverages, feeds, and fermentation processes. Sophisticated enzyme technology now enables starch and its derivatives to be converted to high fructose syrup or used as a substrate for single cell protein (SCP). High fructose syrup is economically competitive with sucrose and offers comparable sweetness. SCP, with its suitable balance of essential amino acids, has potential value for countries with marginal or inadequate animal production capability.

Starch is readily isolated from the roots in the laboratory (31). Washed roots are ground in water to a slurry
which is filtered through a 150 mesh screen and subsequently
through fine muslin. The starch is allowed to settle from
the slurry and centrifuged to separate fiber. It is collected and air-dried. The yield varies but amounts representing
50-56% of the root dry weight can be obtained. Isolation in
this manner provides a product which is free of cucurbita-
cins, the extremely bitter glycosides which are present in
the roots, leaves and fruit of this plant.

Certain physical and chemical properties of Buffalo
gourd starch and other starches are shown in Table 9.

The Buffalo gourd is compared with other root and tuber
starch sources for moisture and starch content on a freshly
harvested basis in Table 10.

A study of the structural features of Buffalo gourd
starch is being made. Fractionation to provide amylose and
amylopectin has been successful, and the properties of the
linear and branched polymers are being determined. An investigation of the rheological properties of the starch is
also being made.

Foliage Studies. One of the striking features of the Buffalo
gourd is its ability to produce large amounts of foliage.
The prodigious vine growth represents a potential source of
feed for ruminants. Harvesting the vegetation prior to frost

Table 9. Properties of Buffalo gourd starch and other
starches.

Analysis	Buffalo gourd	Potato	Tapioca	Corn
Protein, %	0.85	0.05	0.19	0.83
Lipid, %	0.57	0.50	0.99	0.61
Ash, %	0.10	0.35	0.18	0.07
Phosphorus, %	0.022	0.075	0.01	0.018
Granule diam., mean, μm	6	50	17	18
Granule diam., range, μm	2-17	15-100	5-35	10-25
Gelatinization temp., °C	57.0-60.5	58.0-66.8	58.5-70.0	62.0-70.8
Amylose by blue value, %	26.0	24.0	20.0	20.3
Iodine affinity	5.06	4.61	4.43	4.02

Table 10. Moisture and starch content of certain starch
sources.

Source	Starch designation	Moisture	Starch
		(%)	(%)
Solanum tuberosum	Potato	75-78	19
Manihot utilissima	Tapioca	60-75	12-33
Maranta arundinaceae	Arrowroot	65-75	22-28[1]
Cucurbita foetidissima	Gourd root	68-72	15-17

[1] Industrial yield typically 15% due to the dense cell walls
surrounding the starch.

exposure gives a material which is nearly 60% digestible with
a crude protein level in the range 10-13% (dry weight basis).
Experimental diets based on such material have a digestible
energy estimated at 2 kcal/kg dry matter (32). Conversion to
silage offers another unexplored method of utilizing this
material.

In summary, the Buffalo gourd has the potential of becoming a cultivated food and feed crop. It is one of the few
xerophytic plants of possible value which are undergoing significant genetic selection and improvement (33) for arid
lands agriculture. It has been recognized as one of 36
plants which show "promise for improving the quality of life
in tropical areas" (34). It is unusual in producing edible
oil, protein, and starch, all in amounts sufficient to give
it promise as a new food and feed crop for arid lands.

Acknowledgments

The authors wish to acknowledge Gary Nabhan and Joe
Scheerens for reading the manuscript and Madonna Brooks for
typing the manuscript. The authors acknowledge financial
support for the Buffalo gourd research to the N.S.F. (grant
AER-76-82387), the Herman Frasch Foundation, and the
Arizona Agricultural Experiment Station, Acknowledgement is
also extended to Dr. J.T. Lawhon, Texas A&M University, who
conducted small scale processing trials on Buffalo gourd
seed.

References

1. H.C. Cutler, T.W. Whitaker, Amer. Antiquity 26, 469 (1961).
2. L.H. Bailey, Gentes. Herb. 6, 265 (1943).
3. W.P. Bemis, A.M. Rhodes, T.W. Whitaker, S.G. Carmer, Amer. Jour. Bot. 57, 404 (1970).
4. W.P. Bemis, T.W. Whitaker, Madroño 20,33 (1969).
5. I. Grebenscikov, Zuchter 28, 233 (1958).
6. W.P. Bemis, J.M. Nelson, Jour. Ariz. Acad. Sci. 2, 104 (1963).
7. W.P. Bemis, Jour. Amer. Soc. Hort. Sci. 95, 529 (1970).
8. H.J. Dittmer, B.P. Talley, Bot. Gaz. 125, 121 (1964).
9. L.C. Curtis, Chemurgic Digest 13, 221 (1946).
10. H. Shahani, D.K. Markley, J. Quinby, Jour. Amer. Oil Chem. Soc. 28, 90 (1951).
11. S. Paur, N.M./Press. Bul. 1064 (1952).
12. L.C. Curtis, N. Rebeiz, Ford Foundation Prog. Rept. (1974).
13. C.W. Weber, W.P. Bemis, J. Berry, A. Deutschman, B.L. Reid, Proc. Soc. Exp. Biol. Med. 30, 761 (1969).
14. T.P. Hensarling, T.I. Jacks, A.N. Booth, Jour. Agric. Food Chem. 21, 986 (1973).
15. W.P. Bemis, L.C. Curtis, C.W. Weber, J.W. Berry, J.M. Nelson, A.I.D. Tech. Series. Bul. 15, (1975).
16. W.P. Bemis, J.W. Berry, C.W. Weber, T.W. Whitaker, HortSci. 13, 235 (1978).
17. J.C. Scheerens, W.P. Bemis, M.L. Dreher, J.W. Berry, Jour. Amer. Oil Chem. Soc. 55, 523 (1978).
18. P.D. Hurd, E.G. Linsley, T.W. Whitaker, Evolution 25, 218 (1971).
19. J.W. Berry, J.C. Scheerens, W.P. Bemis, Jour. Agric. Food Chem. 26, 354 (1978).
20. J.W. Berry, C.W. Weber, M.L. Dreher, W.P. Bemis, Jour. Food Sci. 41, 465 (1976).
21. W.P. Bemis, J.W. Berry, C.W. Weber, Conf. Proc. Non-conventional Proteins and Foods, Univ. Wisc. 77, (1977).
22. H.K. Goering, P.J. Van Soest, Agricultural Handbook No. 379, Agricultural Research Service, United States Department of Agriculture (1970).
23. M.L. Dreher, M.S. Thesis, University of Arizona (1976).
24. S.A. Thompson, C.W. Weber, J.W. Berry, W.P. Bemis, Nutr. Reps. Int. (1978) In press.
25. J.T. Lawhon, Personal Communication (1978).
26. I.E. Liener, Jour. Food Sci. 41, 1976 (1976).
27. D.S. Bolley, R.H. McCormack, L.C. Curtis, Jour. Amer. Oil Chem. Soc. 29, 470 (1952).
28. B.F. Harland, D. Oberleas, Cer. Chem. 54, 827 (1977).
29. T.J. Jacks, T.D. Hensarling, L.Y. Yatsu, Econ. Bot 26, 135 (1972).

30. P. Girgis, F. Said, Jour. Sci. Food Agric., 19, 615 (1968).

31. J.W. Berry, W.P. Bemis, C.W. Weber, T. Philip, Jour. Agric. Food Chem. 23, 825 (1975).

32. L.B. Waymack, C.W. Weber, J.C. Scheerens, Arizona Cattle Feeders' Day, University of Arizona 9-1 (1976).

33. R.S. Felger, G.P. Nabhan, Ceres/FAO, Review on Development 9, 34 (1976).

34. National Academy of Sciences/National Research Council, "Underexploited tropical plants with promising economic value", 188 p. (1975).

5

Mesquite

An All-purpose Leguminous Arid Land Tree

Peter Felker

Abstract

Managed orchards of selected varieties of mesquite
(Prosopis glandulosa and related species) offers an unexploited potential as a low water, nitrogen, and tillage
requiring protein and carbohydrate source for humans and/
or livestock. Mesquite pods which contain approximately
13% protein and which may contain up to 30% sucrose,
were the major food staple for the Indians in the southern
California deserts.

An orchard of mesquite trees would obviate the need
for tillage and prevent wind and water soil erosion.
Since mesquite's root nodules convert atmospheric nitrogen
into a form suitable for plant use, nitrogenous fertilizers from fossil fuels would not be required. Mesquite's
woody habit and long tap root, which can often reach ground
water at depths of 20 m, confer an unusually exceptional
drought resistance. After trees pass their age of maximum
pod productivity, the wood may be used as a fuel source.

The literature suggests that mature mesquite orchards
receiving no irrigation or nitrogen after establishment,
may yield 4,000 to 10,000 kg/ha in regions where there is
groundwater or 250-500 mm annual rainfall.

Introduction

In North America mesquite is the common name for
perennial woody legumes of the genus Prosopis. Mesquite
may exist either in the form of a 2-m tall shrub (1) or as
a 15-meter tall tree with a 1-meter diameter trunk (2).

Fig. 1.   Mesquite tree with pods near Palm Springs,
          California.

There are 43 species of mesquite indigenous to North and South America, Africa, and Asia (2) many of which naturally hybridize (3, 4).

All the Prosopis species have indehiscent pods which may be tightly coiled or straight and may vary in length from 3 to 30 centimeters (Fig. 2). The pods are about 13% protein (5) and may contain up to 30% sucrose. The seeds vary in length from 0.2 to 0.8 cm. The entire seeds contain approximately 27% protein (5, 6) but after removal of the thick seed coat they contain 55-69% protein (6, 8). Median Prosopis pod and foliage yields have each been reported to reach a maximum of approximately 7,000 Kg/ha/yr when growing in a Chilean salt desert existing solely on ground water (7). The use of Prosopis pods as a food staple for Indians has been the subject of excellent reviews in North America (9), in South America (10), and has been discussed for the India-Pakistan region in numerous papers (11, 12, 13). Mesquite has also provided an important source of wood for building materials and heating and cooking fuel in North America (9, 14, 15), South America (10, 16) and India (11, 12, 13).

If ground water is available within 10 meters of the surface, such as may occur along dry stream courses (17), mesquite will grow in regions such as Death Valley, California where the daily average maximum temperature for July is close to 45° C (18), and where the average annual rainfall is as low as 50 mm (19) (Figure 3). Trials designed to test the adaptability of mesquite for reforestation of 520,000 km^2 of West Pakistan with summer temperature over $43^{o}C$ and rainfall under 250 mm have been successful (20). One can conclude that mesquite is drought resistant.

Mesquite has also been shown to nodulate and increase the nitrogen content of potted plants in the greenhouse (21) as well as that of soil under mesquite trees in natural ecosystems (22).

Unfortunately some mesquite are very aggressive and may form stands of over 1,000 trees/hectare to the displeasure of cattle ranchers in southwestern U.S.A. Accordingly, methods for mesquite eradication have been exhaustively studied (23).

The mesquite literature is vast as mesquite is indigenous to four continents (2). Excellent mesquite reviews are available covering taxonomy (2), physiology,

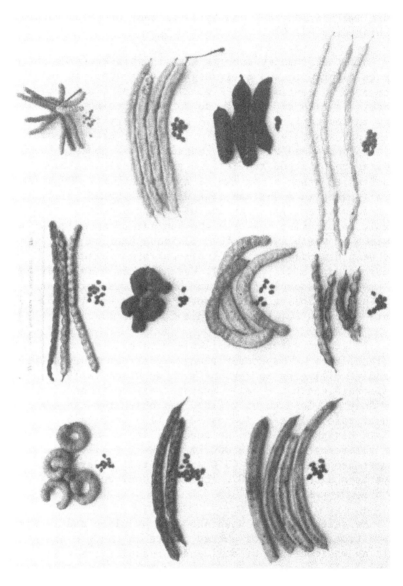

Fig. 2.   Pods of various Prosopis species.
Top (left to right) P. pubescens (California),
P. glandulosa var. torreyana (California), P.
kuntzei (Argentina), P. articulata (Mexico);
2nd row (left to right) P. alba (Argentina),
P. farcta (Near East), P. spp. (Argentina),
P. spp. (New Mexico); Bottom row (left to
right) P. tamarugo (Chile), P. spp. (Texas)
P. pallida (Hawaii).

Fig. 3. Mesquite on sand dunes in Death Valley, California where the highest air temperature in North America of 56°C (134°F) has been recorded.

plant and insect ecology, and ethnobotany (24) and
eradication (23). This chapter will concentrate on those
aspects of mesquite which bear on the development of
mesquite as a useful plant for increasing wood for firewood and building materials, for increasing soil fertility
and for providing livestock and/or human food.

Uses of Tree Legumes
Other than Prosopis

Mesquite appears to have a great potential in semi-
arid regions, but it is only a single example of the way
many genera of leguminous trees and shrubs have bettered
the lives of countless people in semi-arid regions. To
provide a more general perspective of the role of leguminous trees in semi-arid climates, the following overview
will be given. Reviews of this subject are available
(25, 12).

In Australia the Acacia low woodland type scrub
occupies 490 million hectares. Mulga (A. aneura) occurs
as a major component of this woodland scrub (26) and
constitutes a valuable source of livestock feed during
droughts (27). In the dry regions of India and Pakistan,
A. senegal, Prosopis cineraria and North American introduced mesquite (Prosopis spp.) have provided an important
source of heating and cooking fuel and animal fodder
for the local population (12). On the island of Cyprus,
the pods of Carob (Ceratonia siliqua) have provided an
important source of revenue to the local economy (28).

In the areas south of the Sahara, A. senegal has
been an important source of revenue for gum arabic in
regions of 200-400 mm rainfall (29). In regions of 500-
700 mm annual rainfall, A. albida enriches the soil in
farmers' millet and peanut fields (30) and provides pods
for livestock food (31). In savannah areas with moisture
greater than 700 mm annual rainfall, the nere (Parkia
clappertoniana) is a much appreciated tree (32), as its
pods are used for livestock food (33) and to ferment a
product consumed by humans as a condiment with soups (34).

In the semi-arid regions of South Africa, the carob
(Ceratonia siliqua), the honey locust (Gleditsia
triacanthos) and mesquite (Prosopis spp.) have all been
used for animal fodder (35).

In South America Prosopis alba, P. nigra, and P.
chilensis have been very important for building materials
and human and livestock food in the semi-arid region of
northern Argentina (2), as has been P. affinis in Uruguay,
and P. tamarugo in the Atacama salt desert (36, 7).

As discussed by Felger in another chapter of this
book, the leguminous trees ironwood (Olneya tesota),
Palo verde (Cercidium floridium), quaymuchil (Pithecellobium dulce) and mesquite (Prosopis glandulosa and P.
velutina) were a very important food source for humans and
livestock in semi-arid North America.

Although not as well adapted to semi-arid climates as
to climates receiving 600-1700 mm annual rainfall, the
leguminous tree Leucaena leucocephala is especially promising for wood and forage production because of its extremely
rapid growth (37). This genus is also discussed in another
chapter of this book.

The dry forests of the tropics and sub-tropics cover
an area of 31 million sq. km, an area larger than the
U.S.S.R. (23 million sq. km), and in many cases Acacia and
Prosopis are important components of these dry forests
(12). An in depth discussion of all leguminous trees in
these areas would be too lengthy to consider here. Therefore, Prosopis will be used to illustrate the uses,
potential uses, and problems that need to be solved for
leguminous trees in general in semi-arid climates.

Human Use of Mesquite

One of European man's first recorded associations
with mesquite was in 1540 when Alarcon's expedition was
presented with mesquite bread, yucca, rabbits, skins,
fish, and parrot feathers (38).

Writing on the uses of food plants in what is today
southern California, Barrows (15) remarked, "On the desert
the main reliance of the Coahuilla (sic) Indians is the
algaroba or mesquite", and "Everywhere in the Colorado
country, to the Mojave, Yuma, Cocopah, as well as the
Coahuilla (sic), they are the staple of life." Castetter
and Bell (38) reported that "among the tribes on the
Colorado and Gila Rivers mesquite and screwbean pods
constituted the chief source of wild food. Mohave, Yuma,
and Cocopa informants agreed that no other wild food

Fig. 4. Mesquite food preparation implements used by Mrs.
Ruby Modesto's grandparents. In far background
is a mesquite storage basket. Closer can be seen
a "mortar and pestle" used for grinding mesquite
and other grains and in immediate foreground is a
cottonwood mortar most often used for grinding
mesquite pods.

compared in importance with these two; that they were more important than maize (and later wheat); and that they virtually supplied living through the winter until the next crop was ready." For the Seri Indians in Baja, California, mesquite was listed as the most important of over 70 food plants (39).

Mesquite's uses and preparation as a food are very adequately described elsewhere (9, 14). However, for the sake of completeness the Cahuilla Indians' food uses of mesquite will be described here, essentially as outlined by Bean and Saubel (14).

Before the mesquite harvest began, a religious sanction ceremony was necessary in which the ceremonial and political leader picked the first crop. After the mesquite pods were prepared and eaten by members of the lineage group, it was announced that all could go and harvest the mesquite.

For the several week period over which the pods would ripen, families would go to the groves which were inherited by their specific lineage. Only the more productive groves that produced heavy crops on a regular basis were "owned" by specific lineages. Less productive or less reliable trees were utilized on a first-come-first-served basis. All members of the family entered into the mesquite bean harvest with the more agile children collecting the harder to reach beans. Tunnels and "rooms" were made in the thickets to facilitate access to the thickets. Various tools could be left in the "rooms" of the mesquite thickets because the groves were used by only one lineage.

The yellow but immature juicy pods were prepared by pounding or crushing them in a mortar made of a large cottonwood or mesquite stump with a hollowed center (Fig. 4). The resulting pulpy extract was drunk continuously throughout the summer.

The dried mature pods, about 10 to 18 cm long, could be broken into several small pieces and eaten directly. More often they were ground into a meal in a wood or stone mortar. The seeds (which contain a great deal of protein) were most often discarded (Ruby Modesto,* personal communication). The flour, which represents the mesocarp and ectocarp of the pod, was the starting material for numerous food preparations. The meal could be dampened and formed into balls or cakes for storage. Pieces of these cakes were often carried along on various kinds of excursions to provide sustenance.

While dried mesquite meal and cakes were often stored in the home, the largest mesquite quantities were stored as intact beans in elevated circular granaries constructed of woody but pliable long slender shoots of several kinds of plants. These plastered airtight containers might hold 350 to 530 liters (10 to 15 bushels) and were said to be sufficient to feed a family of six to ten people for a year.

It is this author's opinion that the unripe but yellow juicy pods of sweet mesquite trees (and not all mesquite trees are sweet) are similar in taste to immature pea pods known variously to present day Americans as snow peas or sweet peas, while exceptionally sweet pods may taste like bing cherries. The mature dry yellow mesquite pods from sweet trees have an aromatic taste somewhat resembling molasses and the burnt coffee used in cake flavorings.

Very similar mesquite products were used for human food in the semi-arid regions of Argentina, Chile and Peru (16). Among these products is patay, which is a very sweet floury paste obtained by grinding the mature dry pods. After eliminating the seeds and the more fibrous tissue from the patay, it is compressed into special vessels where it stores very well in semi-arid climates. The patay is more valuable as a source of calories than protein as it contains only 4-5% protein and 45% of various sugars.

A kind of honey or mel which is a "concentrated infusion of the fruits" is prepared from special

---

* Mrs. Ruby Modesto is a member of the Torres-Martinez Indian Reservation near Thermal, California and is a member of the A'wilem 'dog' clan lineage of the desert Cahuilla. Mrs. Modesto speaks and teaches her native Cahuilla language and has served as a consultant to many anthropologists and ethnobotanists.

<u>Prosopis</u> varieties (16). Anapa is a sweet, refreshing
drink prepared by crushing the pods in a mortar and
pestle with addition of a little water. Anapa appears to
be analogous to a drink prepared by the Cahuilla Indians
as described above. A fermentation of ground pods of
special white <u>Prosopis</u> is known as aloja and is highly
recommended as a beer or wine substitute by Burkart
(16). Due to the high sugar content and inexpensive
nature of the pods, fermentation and distillation of
ethanol from <u>Prosopis</u> has been suggested. Additional
uses of <u>Prosopis</u> as a source of tannins, gum, and folk
medical remedies have been described. Recipes for the
adaption of these Argentine food and beverage products to
Brazilian customs have been reported (40).

A thorough recent review of <u>Prosopis</u> use in Argentina
(10) describes the tremendous importance of mesquite as a
food source for Indians and the wood as a fuel in the 20th
Century. Here, the wholesale destruction of <u>Prosopis</u>
forests has occurred without reforestation efforts taking
place. This was most severe during the two world wars
when <u>P. caldenia</u> wood was used to fire industrial furnaces
and steam locomotives in the face of a coal shortage.
Because the large trees were harvested first, a negative
selection for tree growth has occurred (10).

While <u>Prosopis</u> wood is hard and has a high tensile
strength, it is not widely used for construction because
long timbers cannot be obtained from the short twisted
trunks. Today the wood is highly prized for parquet floors
and the very hard <u>P. kuntzei</u> wood is used for turnery
articles.

Because of its beneficial characteristics, mesquite
was introduced into Hawaii about 1830, and by 1910, a
Hawaii Agricultural Experiment Station Report labeled it
"the most valuable tree thus far introduced into the
Island" (41). Mesquite grew on the dry side of the
island, where it was estimated during the 1920's that
500,000 bags of dry pods were annually picked and stored
for animal feed (42). In 1926 the Hawaiian Commercial
and Sugar Company hired laborers to pick 900 to 1,250 MT
of <u>Prosopis</u> pods which were ground for animal feed (42).

At the turn of the century, mesquite was introduced into South Africa, where it was used as a fodder tree for the
semi-arid regions of the Transvaal. Loock (43) reported

that ripe crushed mesquite pods had the same nutritional
quality as maize when used as livestock feed. Both Loock
(43) and Jurriaanse (35) discuss cultural requirements and
spacing conditions, and suggested yields for managed
Prosopis stands.

Perhaps, because of mesquites' beneficial qualities in
North and South America, it was introduced into the semi-
arid regions of India by the British in the mid 19th
Century. In India, mesquite has proven a very valuable
plant for providing livestock food, for preventing expan-
sion of the desert, and for providing cooking fuel to the
local population. In Jodhpur Province, mesquite was declared the "Royal Plant" because it supplied the bulk of
the fuel to the local population (13).

A specific example of the use of Prosopis is recorded
for Andhra Pradesh area of India where dry winds caused the
movement of sand which buried agricultural fields, irrigation channels, roads, and even parts of villages (11).
Accordingly the government established a shelter belt with
transplants of Prosopis juliflora, Pithecellobium dulce,
Albizzia lebbek, and Casuarina but later solely used
P. juliflora as it gave the best results in all plantations.
As well as solving the problem of sand invasion of agricultural lands and villages, Prosopis wood solved an
acute fuel shortage, and Prosopis pods greatly relieved
the cattle feed problems in the hot dry areas. A cottage
industry also became established to sell about 2,250 kg
of cleaned Prosopis seed/year to the Forest Department.
Recent reviews covering the use of Prosopis in the reforestation of semi-arid regions are available (44, 29).

Biochemical and Nutritional Properties
of Mesquite Pods

Analyses of entire Prosopis pods have shown them to
contain from 9 to 13% protein, from 13 to 36% sucrose
(45, 6, 5) and from 45 to 55% total carbohydrate (6, 36).
Sodium dodecylsulfate gel electrophoresis of the seed
proteins showed them to have subunits of 77,000 and 96,500
daltons (5). Analysis of Prosopis pod carbohydrates by
gel filtration, nmr, and gas chromatography-mass spec-
trometry of the permethylated sugars was reported (5).

Analysis of Prosopis tamarugo pods and leaves showed
no detectable cyanogenic glucosides and alkaloids but

antitryptic factors and hemagglutinins were found in the
seeds (36).

The protein and amino acid composition of two mesquite
accessions was studied by Felker and Bandurski (8), who
reported the protein content of the true seed of each
species (i.e., minus seed coat) to be 60 and 69%.
These protein values are in good agreement with those
reported by Walton (6). The amino acid analysis indicated
that the essential amino acids valine, threonine, isoleucine, lysine, methionine, cystine, and tryptophan were
all below FAO recommended values for human consumption.
The sulfur amino acids were the most deficient, with
tryptophan being near adequacy in P. chilensis and 45%
below adequacy in P. juliflora (8).

There is a great deal of variability in the pod
structure of mesquite; some species have large seeds with
little sugary pulp (mesocarp) e.g. P. articulata, while
other species such as P. nigra have small seeds with considerable sugary pulp. Since the seeds contain the bulk
of protein in the fruit, the seed/pod-weight ratio could
be a useful criterion for selecting desired protein and
sugar contents of new mesquite varieties. Preliminary
measurements of seed/pod weight ratio of 20 Prosopis
accessions from North and South America yielded values
ranging from 8.4% to 33% (Felker, unpublished).

There have been several animal feeding studies using
mesquite but, with one exception, the studies were carried
out 50 or 60 years ago; consequently, their methods and
presentation of data are somewhat different from those in
common use today.

The earliest feeding study was conducted by Garcia
(46), who used mesquite pods in a feeding trial with pigs
and who made several important observations which are too
often ignored when mesquite pods are used as a livestock
food. First, Garcia noted that the pods must be ground to
secure their full nutritive value since 25% of their weight
is seeds, which would otherwise pass undigested through
the animal's alimentary tract. Since Garcia (46) reported
that the seeds contained 69% of the protein and 60% of the
fat of the pods, substantial nutritional losses would be
incurred if the pods were not ground. Garcia's second
important observation was that "mesquite pods should not
be given as the entire ration on account of their consti-
pating effect but this effect is readily overcome if fed

in connection with alfalfa hay, wheat or some other succulent food." The design of Garcia's feeding trial included
2 lots of 4 pigs each, which were given either 100% ground
maize or a 50% mixture of ground maize and ground mesquite
beans. In addition, both lots were supplied with alfalfa
hay ad libitum. The study showed that mesquite was only
4% less efficient than maize during the first four weeks
in increasing weight of swine, but mesquite was 47% less
efficient on a weight basis after the first four weeks.
A possible explanation for this decrease in efficiency
might be due to the accumulation of heat labile phytohemagglutinins and trypsin inhibitors which are present
in some Prosopis seeds (36) and which are discussed in
detail later.

During World War I, preliminary digestibility studies
were made with Prosopis which is called keawe or algaroba
in Hawaii (47). These authors found that (1) mesquite
meal at 93% dry matter contained 6.0 kg of digestible
protein and 55 kg of digestible carbohydrate per 100 kg
of meal; (2) ground mesquite pods at 83% dry matter, with
seeds removed, contained 2.8 kg of protein; and (3)
ground mesquite seed at 86% dry matter contained 25 kg of
digestible protein and 41 kg of digestible carbohydrate
per 100 kg of ground seeds. These data support Garcia's
(46) findings on the importance of grinding pods to release
seed protein.

Later feeding trials in Hawaii (48) indicated that
mesquite meal was equivalent to pineapple bran in milk-
production trials with cattle and that it was inferior to
barley, but superior to pineapple bran in swine-feeding
trials. In addition, the coefficient of digestibility was
70 and 74% for mesquite-pod meal, and green alfalfa,
respectively, when tested in cattle feeding trials with
Holstein and Hereford steers. The Hawaiian values (48)
for mesquite-bean-meal digestibility are lower than those
reported by Fraps (49) who found the digestion coefficients
of mesquite beans and alfalfa hay to be 90 and 75% respectively. This discrepancy could easily be resolved as there
is no reason to believe that Hawaiian and Texan mesquite
varieties (49) were identical and as the Texan study was
conducted with sheep which have a greater ability to
break and digest hard seeds (1).

As opposed to controlled feeding trials with ground
mesquite, which generally show positive results, uncontrolled grazing of cattle on ranges where mesquite pods
are the predominate or sole source of food may result in

deleterious effects to the cattle. In Hawaii, the death
of cattle has resulted from grazing on pastures containing only mesquite. Adler (45), speculating on the cause
for the mesquite-induced illness, wrote: "A high sugar
diet (30%) low in good forage depresses bacterial multiplication. Digestion of cellulose to available sugar is
not accomplished. Protein synthesis by normal bacteria
is sharply reduced." Very similar findings in Texas were
reported by Dollahite (50).

Sheep feeding trials using Chilean P. tamarugo pods
and leaves as the sole food source showed; that the fruit
has a total digestible nutrient value of 45%, that a mixture of pods and leaves should meet maintenance requirements at voluntary levels of intake, and that all animals
studied had a positive nitrogen balance (51). A more
recent nutritional study on P. tamarugo (36) found the
true digestibility of the seeds to be 70%.

Pak (36) reported trypsin inhibitors and hemagglutinins in mesquite seeds but this is not unusual as these compounds are present in most legume seeds. It is somewhat
surprising that they did not find an increase in digestibility for the heated seed. Since Evans and Bandemer (52)
have found that the nutritional quality of Alaska peas
peaks after about 15 minute of autoclaving and subsequently
decreases (probably due to lysine and methionine degrada-
tion), it is possible that the 2-hr boiling-heat treatment
of Pak et al. (36) may simply be too long. A report cited
by Gohl (53) stating that kiln-dried, ground mesquite pods
are much superior to air-dried, chopped pods supports the
possibility that Pak (36) might have seen higher protein
quality if the pods had been heated for shorter periods
of time. The presence of trypsin and hemagglutinin factors in mesquite might also explain why mesquite beans
were initially comparable to maize in Garcia's (46) feeding trials, but after 4 weeks when these inhibitors had a
chance to build up the mesquite pods were much inferior
to maize.

A contemporary example of use of Prosopis pods to
avoid protein calorie malnutrition (PCM) has been reported in southwest India (54) where several villages
with meager agricultural production in a semi-desert
region had an almost total absence of protein calorie malnutrition. Here both unripe and ripe P. cineraria pods
are consumed by the local population with the dried pods
being stored for later use. The P. cineraria leaves and
pods also provide the major food source for the dairy

cattle. The authors concluded that the "custom of eating
P. cineraria pods seems to be the major factor in prevention of PCM in these areas." This was felt to be especially important for young children who picked up the pods,
ate the pulp, and discarded the seeds while playing in the
fields. The nutritious pulp which the children ate contained 11 grams of protein and 250 Kcal/100 grams.

Mesquite as a Fuel

After mesquite stands have passed their age of maximum productivity, the dead trees may be used for cooking
and heating fuel. Such uses are important in the Jodhpur
Province of India, where Prosopis provides the bulk of
fuel to the local population and hence was declared the
"Royal Plant" (13).

Numerous reforestation trials with 17 species of trees
have been made in the semi-arid regions of India on areas
with rocky outcrops, gravel, and shifting sand dunes. In
these trials, P. juliflora was always found to be one of
the faster growing selections (55). In the same general
region, the average above ground fuel yield for 5 ten year
old trees was 136 kg and 54 kg at sites receiving 395 and
268 mm annual rainfall (56). Several workers (57, 58)
at the Central Arid Zone Research Institute in Jodhpur
India have attempted to develop statistical correlations
for P. cineraria (Syn spicigera) between tree height, crown
diameter, and diameter at breast height with fuel (58)
and pod yield (57). The average fuel yield was 230 kg/tree,
with an 85 to 479 kg range, while the average annual pod
yield was 2.9 kg/tree.

In an evaluation of the fuel potential of various
semi-arid trees in West Pakistan, Ahmed (20) determined the
growth rates of various Prosopis species. Three plots,
ranging in size from 115 m² to 230 m², were located in
areas with 400, 250, and 200 mm annual rainfall. In
the general study area there were reported to be 520,000
km² of land receiving less than 250 mm annual rainfall
with summer air temperatures in excess of 43°C. Seven
Prosopis accessions, including the native P. cineraria
and Texan, Mexican, "Arid" and South American species
(P. nigra) were studied. Dryland forest selections of
Zizyphus, Tamarix, Acacia arabica and A. farnesiana were
also included. The tree densities were very high, in
some cases reaching 11,000 trees/ha. After periods of

12, 14, and 15 years the plots were harvested and total
dry matter was measured. Assuming a specific gravity
of 0.7 (59), the average timber yield (clear bole) for
the Prosopis selections ranged from a low of 13 kg/ha/yr
to a high of 8,000 kg/ha/yr on the 250 mm site and nine
plots had timber yields in excess of 3,000 kg/ha/yr.
In no case did trees other than Prosopis have the highest
fuel yields. It is clear that a tremendous variability
exists in the genus Prosopis for dry matter production,
and that reports of a failure of a single Prosopis accession must not be construed to mean that the site will
not be adaptable to a different Prosopis selection.

Prosopis annual timber volume yields of 4,000 kg/ha/yr
in regions of 250 mm annual rainfall are higher than would
be expected from typical savannah net primary productivity
which would be about 2,000 kg/ha (60). (This assumes the
studies were conducted in an area where a high water table
does not exist which would give the trees an advantage,
although this is not stated.) It is probably safe to
assume that the aerial parts excluding the bole, and the
roots each contribute as much annual growth as the clear
bole. The result is that 4,000 kg/ha/yr clear bole production .is roughly equivalent to a dry matter production of
about 12,000 kg/ha/yr. While there is a report of an
Indian-derived Prosopis savannah as having a net primary
productivity of 14,000 kg/ha (60) at 360 mm annual rainfall, productivity of this magnitude at these low moisture
conditions is truly phenomenal and will require a re-
evaluation of some of the productivity concepts for semi-
arid regions.

Mesquite Water Use Efficiency

Results of a water use study of Arizona range plants
conducted out of doors with pots containing 140 kg of soil
indicated that the shrubs palo verde (Cercidium microphyllum), jojoba (Simmondsia chinensis), mesquite (Prosopis
juliflora), and catclaw (Acacia greggi) were severalfold
more inefficient in their water use than annuals (61).
Mesquite had a water use efficiency of 1730 kg water/kg
dry matter accumulation.

Such a mesquite water use efficiency is in marked
contrast to what can be calculated from ecosystem data.
If one divides Bazilevich and Rodins (62) Prosopis
spp. dominated savannah net primary productivity 14,500

kg/ha/yr by an annual rainfall of 361 mm (3.61 x 10^{+6} kg H_2O/ha) a water use efficiency of 250 kg water/kg net primary productivity results.

If one assumes that the clear bole of Ahmeds' (20) Pakistani Prosopis trees constitutes one half of the above ground dry matter, then dry matter production ranges from 200 kg/ha/yr to 12,000 kg/ha/yr, at annual rainfalls of 394 and 246 mm respectively, and this would yield water use efficiencies ranging from 205 kg H_2O/kg dry matter to 19,700 kg water/kg dry matter. The mesquite water use efficiency measured by McGinnies and Arnold (61) of 1730 kg H_2O/kg dry matter falls within this range.

The range of water use efficiency points out the enormous genetic potential that exists in mesquite and stresses that a water use study on a single Prosopis selection cannot be used to extrapolate water use for the genus Prosopis in general.

Mesquite Ground Water and Salinity Requirements

A thorough study of the relationship between ground water depth and mesquite height was carried out by the U.S. Geological Survey (17) in Southern California where groundwater has much more influence than rain over mesquite growth because of the low annual rainfall (100_omm) and high temperatures (average daily July maximum of 43°C). By studying a transect with a gradient in depth to ground water, it was found that mesquite was only 60 to 90 cm tall when the water table was 14 meters deep, but that mesquite was 3.6 to 6 meters tall when the water table was 3 meters deep.

In a review of literature on use of plants as ground water indictors, Meinzer (17) cites a reference that states "where the depth to ground water is between 7.5 to 9 meters the (mesquite) trees become continuous forests covering the ground" and speculates that water will generally be found under mesquite at depths between 3 to 15 meters. The presence of mesquite roots at the much greater depths of 53 (63) and 80 meters (64) has been reported.

Mesquite's water quality requirements can be inferred from the ground water mineral concentration where mesquite naturally occurs. The chloride plus sulfate concentration in an Arizona site seldom exceeded 100 ppm, while all sites examined near Tularosa, New Mexico exceeded 100 ppm, with one site having a chloride plus sulfate concentration of 6,600 ppm (17). In a Chilean study where pod plus leaf production was 12,000 kg/ha, the ground water salinity ranged from 1,000 to 3,500 ppm (7).

Water with salinity in the range of 1,000 to 3,000 ppm is suitable for irrigation of moderately saline-resistant crops (65), but a chloride plus sulfate ground water concentration of 6,600 ppm which occurred under mesquite in Tularosa, New Mexico would have an electrical conductivity of at least 10 mmhos/cm (65) and would only be suitable for a highly salt tolerant plant (65). These salinities lie within the range found in spent power plant cooling water which in California in the decades to come may require as much fresh water as is now being used for California agriculture (66). It follows that mesquite would be a good candidate for use of spent power plant cooling water.

Nitrogen Fixation

Mesquite seedlings have been shown to nodulate in pot culture using native mesquite soil as inoculum (13, 67, 21, 68), and a positive correlation between plant height, plant weight, weight of nitrogen/plant and number of nodules over 4 mm diameter was observed (21). Nitrogen fixation as assayed by the acetylene assay has been demonstrated in pot culture (Eskew, personal communication). Cross inoculation was observed between Acacia senegal and Prosopis cineraria and between P. glandulosa var. glandulosa and astragalus and Coronilla rhizobium but not between P. cineraria and Acacia tortilis (67).

Strangely enough none of the above authors detected nodules on plants growing in the field. The following explanations have been offered: absence of suitable rhizobia; presence of nodules only after a rain; presence of nodules only in low lying areas where there is adequate soil moisture (69, 21); presence of nodules only on fine lateral roots distant from the tap root (68, 21); presence of nodules at various depths to at least one meter (69); repression of nodule formation by an accumulation of soil nitrogen provided by nitrogen fixation over a long time span (25 years) in mature stands (70) and failure to investigate nodulation at the correct plant developmental stage, e.g., leafing out, flowering, etc.

In spite of a lack of field nodule observations, the increased soil nitrogen and organic matter content under mesquite and other leguminous trees suggests nitrogen fixation occurs. A several-fold increase in soil organic matter and nitrogen content was found under mesquite in Arizona and India (22, 71) and under Acacia albida in Senegal and Sudan (72, 73) while no such increase was found under a non-leguminous control tree Balanites aegyptica (73) or the shrub Guiera senegalensis (74). Additionally, soils taken from under mesquite in Arizona had a four-fold greater herbage yield when grown in growth chamber studies than soils outside of mesquite's foliage cover (22).

The exceedlingly low soil organic matter and nitrogen contents of the previously mentioned soils outside of the influence of leguminous trees (0.3% and 0.03%, respectively) are 10-fold lower than found on temperate agricultural soils. These low values are probably due to the high soil temperature for reasons outlined by Jenny (75). Perhaps the reason leguminous trees occupy such a major part of the dry forests of the world (which cover 31 million square kilometers or an area larger than Russia (12)) is because of their ability to colonize such poor soils. It is not surprising that some workers are coming to the conclusion that plant productivity in semi-arid climates is more limited by soil fertility than by water (76, 77).

While it is difficult to assess the amount of nitrogen fixation by leguminous trees, desert ecosystem productivity data can provide a rough indication of the amount of nitrogen necessary for the ecosystem. A semi-arid ecosystem with a combined nitrogen input from rainwater and blue-green-algal-lichen crusts of 2-3 kg N/ha/yr suggested by Eskew and Ting (68) would provide 10 to 20 kg of protein/ha/yr. At a leaf protein content of 10% this nitrogen input would yield 100 to 200 kg dry matter/ha/yr. If ten years nitrogen could be pooled and recycled without loss, an undisturbed desert ecosystem could have a net primary productivity of 1,000 to 2,000 kg/ha/yr, which is reasonable for desert grassland ecosystems receiving 150-300 mm annual rainfall (60). Therefore, a desert ecosystem in which annual rates of dry matter removal total more than 200 kg/ha/yr probably is being supported by symbiotic nitrogen fixation. If Chilean Prosopis tamarugo pod and leaf yields of 12,000 kg/ha/yr (7) which average 10% protein or 1.6% N (36) are effectively removed from the ecosystem by sheep and sold in the city, a deficit of 200 kg N/ha/yr would occur. Presumably this nitrogen arises from nitrogen fixation.

Not only is nitrogen fixation important for providing the plant its major nutrient, but the resultant increased plant productivity results in increased leaf litter and soil organic matter. In west Africa the increased soil organic matter under the leguminous tree Acacia albida has been shown to provide increased soil cation exchange capacity (or the ability to increase the soil calcium and magnesium concentration), increased soil water holding capacity, and better soil structure (72).

Leguminous trees have a unique advantage over annual legumes in dealing with the inhibitory effect of drought stress on nitrogen fixation because the deep-rooted leguminous trees may reach moisture, and thus relieve the plant of water stress for a longer time in the year than is possible with annuals. For instance a soil water matric potential of -1500 kPa (-15 bars), which is the permanent wilting point, was found to extend from the surface to a depth of 1.4 meters at the end of the dry season in a West African soil receiving 600 mm annual rainfall. In contrast, at a depth of 4 meters the soil water matric potential was -100 kPa (-1 bar), at which almost any plant would not be unduly stressed (30). An illustration of the ability of leguminous trees in semi-arid climates to increase the soil fertility more than annual legumes can be found in West Africa where yields of peanuts are increased if grown beneath Acacia albida trees (30).

Mesquite exists on 30 million hectares of land in the U.S.A. (78) as well as millions of hectares on other continents, but as yet not a single field acetylene assay or other field study of mesquite's nitrogen fixation has been reported in the literature. It seems reasonable that a small research investment to quantify and delimit mesquite's nitrogen contribution to semi-arid soils and to identify a few useful mesquite rhizobia strains would be very profitably spent.

Yield of Mesquite Pods and Leaves

The most comprehensive yield data come from a Chilean study in the Atacama Desert where rain often does not fall for several years but where the ground water is shallow (2 to 10 meters) and somewhat saline (0.1 to 0.35%) (7). In this area the leaves and pods of Prosopis tamarugo are used to support extensive sheep raising operations.

Leaf and pod productivity studies were conducted on 29 plots with trees ranging in age from 9 to 36 years and in spacing from 7 x 7 m to 20 x 20 m (7). Two randomly chosen trees per plot were used in the forage productivity analyses. For trees 6 and 9 years of age, trays of 1 m^2 surface area were placed under the trees at true north and at locations 120° east and west of true north. For trees 12 through 18 years of age, 6 trays were used, and for trees 21 years of age, 9 trays were used. The projected canopy area was measured so that leaf and pod yield could be expressed per tree and per m^2 of canopy cover. The lowest pod yield per tree was 0.19 kg for a 9 yr old tree and the highest was 147 kg of pods for a 32 year old tree. The pod yield per unit of canopy projected cover ranged from 0.008 to 1.45 kg/m^2. Because the tree with the largest pod yield had the widest spacing, the maximum pod yield per hectare (12,700 kg) occurred for trees with 65 kg of pods spaced at 7 m.

Curves were prepared showing canopy closure and pod and leaf yield as a function of tree age. At 30 years of age, the maximum pod plus leaf yield stabilized at 1.4 kg/m^2 with a range of 0.8 to 1.8 kg/m^2. At this age the leaf and pod production were approximately equal.

Assuming a yearly pod and leaf ration of 800 kg per sheep, the sheep carrying capacity was determined to be 12.8 sheep/ha in 20 years for a 7 x 7 m spacing, 14.3 sheep/ha in 30 years for a 10 x 10 spacing, and 13.7 sheep/ha at 36 years for a 13 x 13 m spacing.

Plant productivity of this magnitude (10,000 kg/ha/ yr) is not generally accepted by ecologists at this low rainfall level. Nevertheless the Pakistani data for Prosopis timber volume which ranged from 200 to 8,000 kg/ha/yr occurred at a rainfall of 250 mm (20), and the savannah net primary productivity of 14,000 kg/ha/yr which occurred at 360 mm annual rainfall (62) support the high Chilean Prosopis pod yield data.

Mesquite pod yields from North America have been much less thoroughly studied than in Chile with only three actual pod measurements having been conducted. Garcia (46) harvested the pods from a single mesquite bush in New Mexico to use for pig feeding trials. He obtained 17 kg of pods from a bush located 20 meters above the water table in an area with 150 mm of annual rainfall.

In 1977 we hand picked all the pods from two very different sized mesquite trees on the Torres-Martinez Indian Reservation in the Coachella Valley of southern California. In this region the rainfall is only 100 mm annually but the water table is several meters below the surface. A tree known to be approximately 10 years old, 5 m tall, with a diameter at breast height (DBH) of 5 cm yielded 7.3 kg of pods. A larger tree having two trunks of 25 cm DBH and 8.5 m tall yielded 73 kg of pods with an estimated 10 kg still in the tree. After sun drying on southern California pavement for 3 days, the pod weight was 38 kg, and the total estimated dry pod weight for the tree was 43 kg. In 1978 the larger tree yielded 57 kg of pods which had an oven dry weight of 50.9 kg (Fig. 5). A smaller weight reduction in the 1978 harvest occurred because the pods were harvested at a later stage of maturity when many of the pods had fallen to the gound. This yield would probably have been higher if extensive Psyllid insect damage and parasitic mistletoe were absent from the tree.

In contrast Prosopis average pod yield for 25 tagged trees over a 5-year period was reported to be 14 kg/ha/yr in an area with 280 mm annual rainfall on the Arizona range (78). A yield of 14 kg/ha is clearly at odds with the previous data since a single California tree produced more than an entire hectare in the Arizona study. If the California tree had no production for two consecutive years, it would still have a yield higher than a hectare of Arizona trees, but the California tree has produced pods for the last two years with an average pod yield of 47 kg. Furthermore, a pod yield of 14 kg/ha/yr is almost 500-fold lower than the Prosopis pod yield found in the Chilean study.

D'Antoni and Solbrig (10) have suggested that a negative selection for tree growth has occurred in S. America with the largest trees having been cut first. If a similar event occurred in North America, it might explain the low mesquite productivity on the open ranges and the higher productivity on the uncut stands of the Torres-Martinez Indian Reservation which we measured.

The results of mesquite growth measurements for several North and South American mesquite selections growing under identical conditions on the University of California, Riverside Agricultural Experiment Station are presented in Figure 6. The North American seed material for this trial was collected along an east-west transect

Fig. 5. The pods harvested in 1978 from Mr. and Mrs.
Modesto's mesquite tree near Thermal, California.
The oven dry weight was 50.9 kg (112 lb). A
portion of the tree is excluded to the right.

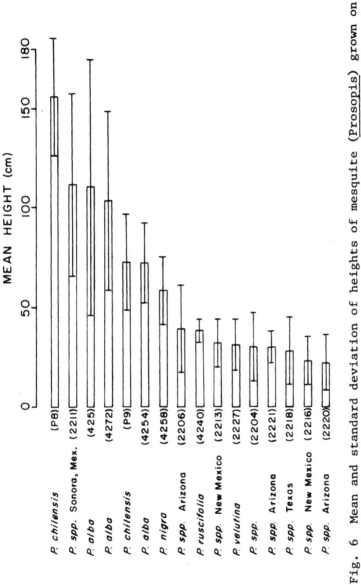

Fig. 6 Mean and standard deviation of heights of mesquite (Prosopis) grown on University of California Riverside Experimental Station. Three-month-old seedlings were transplanted to field 5 July 1977 and measured 22 October 1977. The lines P8, 425, 4272, P9, 4254, 4258, 4240, are of South American origin. All others are of North American origin.

through Texas, New Mexico and Arizona by Simpson* while
the South American seed material was obtained either
through the USDA or from Solbrig** from material collected
during the U.S. International Biological Program (USIBP).
This experiment demonstrates that the growth for South
American mesquite is generally much greater than for the
North American mesquite and suggests the possibility that
a negative selection for growth and perhaps pod yield has
occurred on the U.S. ranges.

In addition to a few quantitative measurements on
mesquite pod productivity, there are many estimates of
mesquite pod yields. Bentley (79) and Smith's (80) estimate of 100 bushel/acre mesquite pod yield is equivalent
to approximately 2,000 kg/ha using a value of 9.5 kg/
bushel (46). If each tree were to yield the 17 kg reported by Garcia, this would yield a density of 118 trees/
ha or a 9 m center-to-center spacing between trees.
Whether trees can survive that close together and maintain
the same productivity at a water-table depth of 20 m is
not known, but much greater densities are achieved on the
Torres Martinez Indian Reservation in Coachella Valley,
California where the water table is only several meters
deep and average rainfall is 100 mm annually.

There are numerous reports of higher pod yields of
mesquite. For example, Wealth of India (81) reported yields
of up to 90 kg/yr for a 10-yr-old tree; Smith (42) reported
yields in Hawaii of between 4,000 and 20,000 kg/ha/yr with
a good forest yielding 8,000 kg/ha/yr; Loock (43) in South
Africa reported that a 10-yr-old tree should yield 90 kg/yr
and suggested a spacing of 6.1 m which would yield over
18,000 kg/ha/yr. The spacing recommended by Jurriaanse
(35) (9 m) and his conservative estimate of 90-140 kg
pods/tree/yr in regions with 250-500 mm annual rainfall
would yield 100 tree/ha with yields between 9,000 and
14,000 kg/ha. In a very sketchy outline, Douglas (82)
reported mesquite yields of 20,000 kg/ha.

This author realizes that mesquite pod yields approach-
ing 10,000 kg/ha/yr under 250-500 mm annual rainfall are an

* Dr. Beryl B. Simpson, Department of Botany,
Smithsonian Institution, Washington, D.C. 20560.

** Dr. Otto T. Solbrig, Department of Biology and
Gray Herbarium, Harvard University, Cambridge,
Massachusetts 02138.

order of magnitude higher than would be expected from
savannah net primary productivity data (60). Nevertheless,
the P. tamarugo pod yield data is soundly established, and
highly productive stands of mesquite have been recorded
with usable timber yields of 8,000 kg/ha/yr at 250 mm rainfall (20) and with net primary productivity of 14,000 kg/ha
at 360 mm rainfall (62). This high level of productivity
coupled wth 1:1 ratio of leaf to pod productivity in mature
P. tamarugo trees further suggests that pod yields approach-
ing 10,000 kg/ha are possible. Furthermore, it is difficult
to believe that Hawaiian (42), South African (35, 43)
and Indian workers (81) are all 10-fold high with their
estimates. It is certain that well spaced, managed,
and especially suited Prosopis strains are required to
achieve this productivity. The Chilean P. tamarugo studies
provide a useful example to follow.

The Mesquite Problem on the Ranges
of Southwestern U.S.A.

The mesquite problem on the ranges of southwestern
U.S.A. stems from the observation that approximately 1%
of cattle existing solely on mesquite pods become ill
(83), some of which may die with a ball of undigested pods
in their rumen (50, 45, 83) and that a pasture will support
more cattle after mesquite removal (23).

The cattle nutritional problem is rather straight-
forward. Some mesquite pods contain up to 30% sucrose
(45) by weight which is very palatable to livestock.
Unfortunately, this amount of sugar appears to repress
bacterial-cellulase activity so that cattle cannot
digest cellulose. The result is that cattle may die with
a ball of impacted pods in their rumen (50, 45, 83).

A second nutritional problem arises in cattle because
the hard coated, high protein seeds often pass undigested
through the rumen of cattle. As mentioned above, mesquite
pods should be ground to secure their full nutritive value.
These difficulties are amenable to correction by planting
mesquite as an orchard crop so that the pods can be harvested, ground and mixed with other rations to lower the
total sugar content. Probably because the sheep chewing
action is more effective than cattle in cracking mesquite
seeds, the unground pods can be fed directly to sheep

which can digest 90% of the nitrogen in mesquite pods
(49). Experiments showed that 69% of the mesquite seeds
passed undigested through yearling steers while 14% of
the seeds passed undigested through the digestive tract
of ewes (1). The identification and use of range animals
especially suited to cracking and digesting hard mesquite
seeds would undoubtedly increase the range productivity,
and decrease the spread of unwanted mesquite trees.

There are at least two reasons why a pasture will
support more cattle after mesquite removal than before.
The first is the finding of Parker and Martin (78) and
Cable (84) that mesquite obtains at least part of its water
from the zone where grasses also obtain their water.
The second is the finding of Tiedemann and Klemmendson
(22) that the organic matter and nitrogen content of
soils directly beneath mesquite foliage cover is several-
fold greater than areas without mesquite cover. The
average herbage yield of grasses grown in a growth chamber
on soils originating directly under mesquite foliage
cover was four times greater than on soils away from
mesquite foliage cover (22). Amendments with nitrogen,
phosphorus, potassium and sulfur showed the predominant
nutrient supplied in the mesquite soil was nitrogen.
Since mesquite is a legume and has been demonstrated
to fix nitrogen (see above), it seems probable that this
increase in soil nitrogen is due to nitrogen fixation,
translocation of the nitrogen to the leaves, and leaf
decomposition in the soil.

It seems reasonable that the increase in grass and
cattle production after mesquite removal is caused by both
an increase in soil nitrogen, probably due to nitrogen
fixation, and a removal of mesquite's competition for
water. More recent data of Tiedemann and Klemmendson (85)
suggest that release of moisture competition by mesquite
may not be as important in stimulating perennial grass
production as previously thought.

It is possible that mesquite's deep roots may provide
a mechanism for overcoming the inhibitory effects of
drought stress on nitrogen fixation which is common in
annual legumes. Perhaps a 20-year-old stand of mesquite
may be viewed as providing a green legume fallow analogous to clover and alfalfa fallows in more mesic regions.

Mesquite thickets 450 meters long and 100 meters wide
have been observed by this author in southern California,
which are virtually impenetrable by man. Such thickets

would be less desirable than isolated trees if the pods
were to be collected and used for food. In Hawaii, thinning of such thickets was reported to increase pod yield
(41). The wood obtained from the thinning of these Hawaiian
thickets was used for charcoal, and almost paid for the
thinning costs. Thickets might be desirable in a program
solely designed for maximum production of woody biomass
or as a legume fallow for soil enrichment. If it is
deemed desirable to thin or eliminate such thickets,
the experience developed in Texas employing chemical and
mechanical thinning methods will be very useful (1, 23).

If maximum pod production is most important from
mesquite stands, a concerted effort will have to be made
to establish and maintain isolated high yielding mesquite
trees in densities between 40 and 200/hectare in a
defined pattern amenable to harvesting pods from the
ground after they have fallen. Such managed, orchard-
like stands will require use of mesquite species such as
Prosopis affinis, P. alba and P. nigra which Burkart (2)
states never become invaders.

Mesquite pod yields of 14 kg/ha reported by Parker
and Martin (78) are dramatically lower than the 7,000 kg/ha
median P. tamarugo pod yields reported in Chile for 30
year old trees. The average 8 year grass forage yield of
300 kg/ha reported by Parker and Martin for mesquite
eliminated grassland is very much lower than mesquite
productivity reported by Baselvich and Rodin (62) and
that which can be calculated from Ahmeds' (20) mesquite
forestry data in Pakistan. Mesquite's growth rates on the
University of California Experiment Station have demonstrated that growth rates, productivity, and, by analogy,
probably nitrogen fixation, are at least four to five
fold lower for most mesquite lines derived from Texas, New
Mexico and Arizona than for South American derived P.
chilensis and P. alba species. Indeed the commercial
desert landscape nurseries in Phoenix and Tucson market
more of the South American derived Prosopis species than
the native mesquite because of the latter's slow growth
rate (Felger, personal communication).

We would agree that the native mesquite pod productivity of 14 kg/ha and forage yields in mesquite dominated
savannah of 300 kg/ha are unacceptable, but it is this
author's belief that the best solution to the problem lies
in the introduction of new mesquite lines with vastly
higher nitrogen fixation, and pod and leaf productivity.

Fig. 7. An introduced, approximately 30-yr.-old
         Prosopis from South America growing near an
         irrigation line in Indio, California. The
         height of the man is 1.62 m.

Genetic Resources, Breeding Structure, and Ecotypic Variation

Johnston (4) has described 6 species of Prosopis with 2 subspecies for P. glandulosa in North America and Hawaii. Burkart (2) recognizes 8 native species in the Texas-Mexico area, 34 native species in South America, 1 native species in tropical Africa, and 3 native species in Southwest Asia, and North Africa. Morphologically, the species of Prosopis vary greatly with some Prosopis rarely being arborescent, e.g., P. glandulosa var. prostrata (2) while P. alba and P. nigra have a typical arborescent habit reaching heights of 12 m with a diameter of 1 m (2) (Fig. 7). Six mesquite growth forms which are dependent on soil and climatic conditions have been described by Fisher et al. (1).

Mesquite leaves show marked variation with P. alba having 20 to 50 leaflets (0.5 to 1.7 cm long and 1-2 mm wide) per pinnae, while P. ruscifola has only 2 much larger (4-10 cm long, and 0.7 to 2.5 cm wide) leaflets per pinna. In contrast, P. kuntzei has no leaves but only photosynthetically active stems (2). See Fig. 8.

A further example of Prosopis genetic diversity occurs in the wood density. Wiley and Manwiler (59) report a wood specific gravity of 0.7 for Texan mesquite, while Burkart (2) reports a wood specific gravity of 1.2-1.35 for P. kuntzei, thus making it considerably denser than water.

A great deal of ecotypic variation in time of budburst, growth, temperature requirements for growth, and frost tolerance have been observed in field and greenhouse studies (86, 87) for North American and two South American Prosopis. As might be expected, greater frost tolerance, diminished height, complete loss of leaves, and dormancy at time of first fall freeze were observed for the most northern ecotypes from Oklahoma, and north Texas (86).

Mesquite's frost tolerance ranges from species which are extremely frost sensitive (86) to the frost hardy P. denudans whose natural habitat nearly reaches 48°S latitude in Argentina (2). A weather station at Santa Cruz, Argentina (50°S Lat.) records an average daily maximum in July of 5°C and an average daily minimum of -2.2°C (88). Such a climate is roughly comparable to that of Cheyenne, Wyoming (88).

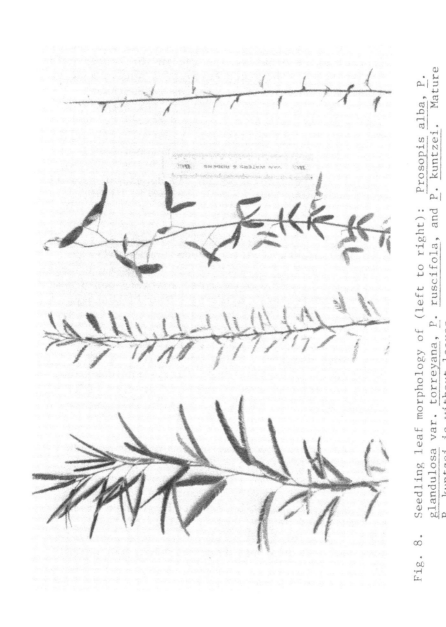

Fig. 8. Seedling leaf morphology of (left to right): Prosopis alba, P. glandulosa var. torreyana, P. ruscifola, and P. kuntzei. Mature P. kuntzei is without leaves.

Outcrossing, interspecific hybridization, and introgression in Prosopis have been reported by several authors (89, 4, 2). Specific examples of putative Prosopis hybrids based on morphological grounds are P. articulata x P. glandulosa (4), P. tamarugo x P. strombulifera, P. ruscifola x P. alba var panta, and P. hassleri x P. fiebrigii (2, p. 468, 477, 479). Cytogenetic evidence for the naturally occurring hybrids P. alba x P. nigra, P. hassleri x P. ruscifola and P. ruscifola x P. alba has been compared with morphological characteristics (3). This outcrossing could be expected because mesquite flowers are hermaphroditic, insect-pollinated (2) and self-incompatible (24). Therefore, one would expect a high degree of outcrossing and heterozygosity. In the short run, mesquite's breeding mechanism will cause difficulties in a breeding program as seeds will not be identical to the parent, but in the long run the ability to outcross will facilitate transfer of desirable genetic characteristics between species.

When mesquite seeds from the same tree were planted in a University of California field study, the parent's heterozygosity was manifest as (1) trees with and without spines, (2) trees which were either single stemmed or multistemmed, (3) trees whose height growth varied by a factor of three.

The failure of mesquite seeds to be identical to the parent will require development of vegetative propagation techniques. A report of mesquite propagation by root and shoot cuttings has appeared (90), but unfortunately this method is really only a severe seedling pruning of roots and shoots designed to enhance viability of transplants in semi-arid regions. The widely used ornamental "Chilean mesquite" has been rooted from semi-hard wood cuttings using a 5 sec - 3,000 ppm indolebutyric acid in vodka (c.a. 50% ethanol) dip with a 1 to 1 perlite/vermiculite rooting medium (Ron Gass, pers. comm.*). We are presently involved with numerous hormone formulations, rooting mixtures and soil temperatures in an attempt to root mesquite cuttings.

As Burkart (2) reports: "There is a real need for more ecological studies on these species (Prosopis) and for a selection program to aid forestation programs in arid and semi-arid regions. Superior strains, especially of such useful species as P. juliflora, P. chilensis, P. pallida, P. caldenia, P. alba, P. nigra, P. affinis, and

---

* Ron Gass, Mountain State Wholesale, P.O. Box 33982, Phoenix, AZ 85067

P. tamarugo, should be cultivated and tested at experiment
stations of dry-climate countries in order to have them
ready for large forestation programs."

Cultural Conditions

The temperature requirements for mesquite growth are
rather high. Hull (91) has reported that maximal shoot
growth occurs at air temperatures greater than $30^{\circ}C$ while
Scifre (23) reports maximal shoot growth at soil temperatures from 29° to $32^{\circ}C$.

The results of one year's field trials on the
University of California Riverside Experiment Station with
selections of Prosopis alba, P. chilensis, P. glandulosa,
P. nigra, P. ruscifola, and P. velutina indicate that the
transition temperature for cessation of growth in the fall
and active growth in the spring occurs when the 2 week
average daily maximum air temperature is approximately
$27^{\circ}C$ $(81^{\circ}F)$.

Before planting mesquite the seeds should be
extracted from the pods and scarified. A motorized device
which eliminates the very laborious hand dissection of seeds
from pods has been described (92). Mechanical scarification is far more effective in breaking seed dormancy than
either boiling water or acid tratment and a simple method
that yielded 95% germination within one week was to shake
450 g of mesquite seeds 2 times/sec for 15 minutes in a
square tin (93). Relatively inexpensive motorized seed
scarifiers are also available (Seedburo Equipment Co.,
Chicago, Illinois).

Most scarified Prosopis seed readily germinate (24
hrs) in filter paper lined petri dishes at about $28^{\circ}C$.
However, recalcitrant seeds such as freshly harvested P.
africana seeds germinate much more quickly in coarse
vermiculite filled 10 cm tall pots receiving $30^{\circ}C$ bottom
heat and a 2 mM gibberellic acid drench.

Spider mites can be a severe problem on some mesquite
accessions in the greenhouse. The normal insecticide for
use on spider mites, chlorbenzilate, was found to severely
defoliate young mesquite seedlings, but pyrethrin based
insecticides and the systemic orthene are not phytotoxic to
mesquite and adequately control the insects.

As thoroughly reviewed by Kingsolver et al. (94),
bruchid insects often may destroy the great majority of
the seeds within mesquite pods. This can be detrimental
if the pods are to be used either as a food or seed source.
While we have verified that fumigation with methyl bromide
or phostoxin or freezing the pods for several weeks at
-20 C kills all the bruchids, we sought a method to prevent
the bruchids from entering the unripe pods still on the
tree. We found the use of orthene spray 3 weeks after
flowering and at subsequent 3 week intervals to be effective in reducing the number of bruchid emergence holes from
23 holes/100 pods on the control tree to 1 bruchid hole/100
pods on the sprayed tree.

Fortuitously the use of orthene spray for bruchid
control led to the discovery of severe Psyllid insect
damage on mesquite since several weeks after spraying with
orthene the mesquite tree initiated new terminal shoots
and flowered while the control tree remained the same.
Upon closer examination it was found that Psyllid insects*
had destroyed the terminal shoots on the mesquite and the
orthene spray temporarily relieved this damage. We have
subsequently observed this problem in Thermal, California
where it has been controlled by malathion spray. Psyllids
are also a problem in Arizona where they have been controlled with diazinon (Sacamano, personal communication**).
How widespread this previously unreported insect pest
on mesquite exists is not known.

Conclusions

The available literature suggests that Prosopis pod
yields from 4,000 to 10,000 kg/ha are possible from mature
trees receiving neither nitrogen nor irrigation after stand
establishment in areas with ground water or 250 to 500 mm
annual rainfall. These yields will require utilization of
non-invasive trees especially selected for pod yield, and
the management of the trees in an orchard type setting with
densities ranging from 50 to 200 trees/ha.

---

* Identification by Dr. Leland Brown, UCR Entomology Dept.

** Dr. C. Sacamano, Dept. of Plant Sciences, Univ. Arizona,
Tucson 85721.

To maximize their use pods must be collected and ground to release the seed protein, mixed with other rations to lower the sucrose concentration, and heated to destroy phytohemagglutinins and trypsin inhibitors if they are to be fed to non-ruminants. Alternatively a range animal better adapted to mesquite's hard seed coat could be used.

The genus Prosopis to which mesquite belongs and forms interspecific hybrids is extremely variable in growth habit, growth rate, pod morphology, leaf morphology and site adaptability to frost, salinity and heat. Different Prosopis selections can and should be developed for various ecological niches, e.g. drought, salinity, and frost and for various uses, e.g. high sugar pods, high protein pods, woody habit for biomass production and for ornamental uses. Failure of one mesquite selection to fulfill a particular use or ecological niche must not be construed to mean that no mesquite is so adapted.

Mesquite should properly be viewed and managed as a long term, multiple use, renewable resource with benefits to be accrued from pods to be used for livestock or human food, from woody biomass to be used as a source of fuel, and as one of the few legumes capable of fixing nitrogen in semi-arid soils and thus increasing the fertility of these soils. There is much room for improvement on the 30 million hectares where mesquite now grows in the United States as well as the millions of hectares in developing countries.

References and Notes

1. C. E. Fisher, C. H. Meadors, R. Behrens, E. D. Robinson, P. T. Marion and H. L. Morton. Control of mesquite on grazing lands. Tex. Agric. Exp. Sta. B 935, 24 pages (1959).

2. A. Burkart. A monograph of the genus Prosopis (Leguminosae subfam. Mimosoideae). J. Arnold, Arb. 57(3): 217 and 57(4): 450 (1976).

3. J. H. Hunziker, L. Poggio, C. A. Naranjo, R. A. Palacios, and A. B. Andrada. Cytogenetics of some species and natural hybrids in Prosopis (Leguminosae). Can. J. Genet. Cytol. 17: 253 (1975).

4. M. C. Johnston. The North American mesquite Prosopis sect. algarobia (Leguminosae). Brittonia 14, 72 (1962).

5. A. A. Figueiredo. Lebensmittelchemische relevante inhaltstoffe der schoten der algarobeira (Prosopis juliflora DC). Ph.D. thesis. Wurzburg, Germany (1975).

6. G. P. Walton. A chemical and structural study of mesquite, carob, honey locust beans. USDA Dept. Bull. 1194: 19 pages (1923).

7. H. E. Salinas and S. C. Sanchez. Informe Tecnico No. 38, Instituto Forestal, Seccion Silvicultura, Santiago, Chile. 35 pages (1971).

8. P. Felker and R. S. Bandurski. Protein and amino acid composition of tree legume seeds. J. Sci. Fd. Agric. 28, 791 (1977).

9. R. S. Felger. In Mesquite -- Its Biology in Two Desert Ecosystems, B. B. Simpson (ed.). Dowden, Hutchinson & Ross Inc., Stroudsburg, Pa. 150 (1977).

10. H. L. D'Antoni and O. T. Solbrig. In Mesquite -- Its biology in two desert ecosystems, B. B. Simpson (ed.). Dowden, Hutchinson & Ross Inc., Stroudsburg, Pa. 189 (1977).

11. P. S. Rao. Shelterbelt plantations of Prosopis juliflora on the Hagari riverbanks in Kurnool District. Indian Forester (90) 658 (1964).

12. A. L. Griffith. In Acacia and Prosopis in the dry forests of the tropics. FAO Rome, Italy. Mimeo 149 pages (1961).

13. R. K. Gupta and G. S. Balera. Comparative studies on the germination, growth, and seedling biomass of two promising exotics in the Rajasthan desert. Indian Forester, 280 (1972).

14. L. J. Bean and K. S. Saubel. In Temalpakh. Malki Museum Inc., Banning, Ca. (1972).

15. D. P. Barrows. In Ethnobotany of the Coahuilla Indians. Malki Museum Press, Banning, Ca. (1967).

16. A. Burkart. Las leguminosas Argentinas. Acme Agency Publishers. Buenos Aires, Argentina (1943).

17. O. E. Meinzer. Plants as indicators of groundwater. U.S. Geological Survey, Water Supply Paper 577, 95 pages (1927).

18. H. A. Mooney. In Mesquite -- Its biology in two desert ecosystems, B. B. Simpson (ed.). Dowden, Hutchinson & Ross Inc., Stroudsburg, Pa., 30 (1977).

19. H. A. Mooney, O. Bjorkman and J. Berry. Photosynthetic adaptations to high temperature. In Environmental Physiology of Desert Organisms, N. Hadley (ed.) 139. Dowden, Hutchinson & Ross Inc., Stroudsburg, Pa. (1975).

20. G. Ahmed. Evaluation of dry zone afforestation plots. Pakistan J. Forestry, 168 (1961).

21. A. W. Bailey. Nitrogen fixation in honey mesquite seedlings. J. Range Man. 29(6): 479 (1976).

22. A. R. Tiedemann and J. O. Klemmedson. Nutrient availability in desert grassland soils under mesquite (Prosopis juliflora) trees and adjacent open areas. Soil Sci. Soc. Amer. Proc. 37, 107 (1973).

23. C. J. Scifres. Mesquite Research Monograph 1. Texas Agric. Exp. Station, Texas A & M Univ. (1973).

24. B. B. Simpson. In Mesquite -- Its biology in two desert ecosystems. Dowden, Hutchinson & Ross Inc., Stroudsburg, Pa. 96 (1977).

25. P. Felker and R. S. Bandurski. Uses and potential uses of leguminous trees for minimal energy input agriculture. Econ. Bot. (in press), (1978).

26. I. F. Beale. Tree density effects on yields of herbage and tree components in South West Queensland Mulga (Acacia aneura) scrub. Trop. Grasslands Vol. 7, No. 1, 135 (1973).

27. M. C. Weller. Mulga as a drought feed for cattle. Queensland Agric. J., 530 (1974).

28. W. N. L. Davies. The carob tree and its importance in the agricultural economy of Cyprus. Econ. Bot. 460 (1970).

29. Anon. In Tree Planting Practices in African Savannas. FAO Rome, Italy, 42 (1974).

30. C. Dancette and J. F. Poulain. Influence of Acacia albida on pedoclimatic factors and crop yields. African Soils, 14, 143 (1969).

31. G. E. Wickens. A study of Acacia albida Del (Mimosoideae). Kew Bulletin Vol. 23, No. 2, 181 (1969).

32. F. Busson, P. Jaeger, P. Lunven and M. Pinta. In Plantes alimentaires de l'Ouest Africain. Inter Agency Publication (Ministere de la Cooperation, Ministerie d'Etat charge de la Recherche Scientifique et Technique Ministere des Armees). Marseille, France, 272 (1965).

33. B. L. Fetuga, G. M. Babatunde and V. A. Oyenuga. Protein quality of some unusual protein foodstuffs: Studies on the African locust bean seed (Parkia filicoidea Welw.). Br. J. Nutr. 32, 27 (1974).

34. B. S. Platt. Improvement of the nutritive value of foods and dietary regimes by biological agencies. Fd. Tech. 18, 662, (1964).

35. A. Jurriaanse. In Are they fodder trees? Pamphlet 116 of Forestry Dept. Private Bag X93, Pretoria, Transvaal, S. Africa, 32 pages (1973).

36. N. Pak, N. Araya, R. Villalon, and M. A. Tagle. Analytical study of tamarugo (Prosopis tamarugo),

an autochthonous Chilean feed. J. Sci. Fd. Agric. 28, 59 (1977).

37. N. Vietmeyer and B. Cotton. In Leucaena: Promising forage and tree crop for the tropics. National Acad. Sci., Washington, D.C. (1977).

38. E. F. Castetter and W. H. Bell. In Yuma Indian Agriculture. Univ. of N. Mexico Press, Albuquerque, N.M., 179 (1951).

39. R. S. Felger and M. B. Moser. Seri Indian food plants: desert subsistence without agriculture. Ecol. Fd. Nutr. 5, 13 (1976).

40. G. Azevedo. Vagens da Algarobeira na alimentacao humana. Mundo Agricola 15, 53 (1966).

41. E. V. Wilcox. The algaroba in Hawaii. Hawaii Agric. Exp. Sta. Press Bull. 26: 8 pages (1910).

42. J. R. Smith. In Tree Crops -- A Permanent Agriculture. Devin-Adair Publ. Co., New York, N.Y. (1953).

43. E. E. M. Loocke. Three useful leguminous fodder trees. Fmg., South Africa. 2:7 (1947).

44. R. N. Kaul. In Afforestation in arid zones. Dr. W. Junk, N. V. Publ. The Hague. (1970).

45. A. E. Adler. Indigestion from an unbalanced Kiawe (Mesquite) bean diet. Amer. Vet. Med. Assn. J., 155: 263 (1949).

46. F. Garcia. Mesquite beans for pig feeding. N. Mex. Agric. Exp. Sta. 28th Ann. Rpt., 77 (1916).

47. M. O. Johnson and K. A. Ching. Composition and digestibility of feeding stuffs grown in Hawaii. Hawaii Agric. Exp. Sta. Press Bull. 53: 26 pages (1918).

48. Anon. Report of the Hawaii Agric. Exp. Sta., Honolulu, Hawaii, pp. 66, 76, 77 (1937).

49. G. S. Fraps. Digestion experiments with oat by-products and other feeds. Bulletin 315, Texas Agric. Exp. Sta. (1924).

50. J. W. Dollahite. Management of the disease produced in cattle on an unbalanced diet of mesquite beans. Southwestern Vet. 17, 293 (1964).

51. L. Latrille, X. Garcia, J. G. Robb, and M. Ronning. Digestible nutrient and nitrogen balance studies on tamarugo (Prosopis tamarugo Phil) forage. J. Anim. Sci. Vol. 33, No. 3, 667 (1971).

52. R. J. Evans and S. L. Bandemer. Nutritive value of legume seed proteins. Agric. Fd. Chem. 153: 439 (1967).

53. B. Gohl. In Tropical Feed. FAO Publications, Rome, 209 (1975).

54. M. C. Gupta, B. M. Gandhi and B. N. Tandon. An unconventional legume -- Prosopis cineraria. Am. J. Clin. Nutr. 27, 1035 (1974).

55. C. P. Bhimaya, R. N. Kaul, B. N. Ganguli, I. S. Tyagi, M. D. Choudhary, and R. Subbayyan: Species suitable for afforestation of different arid habitats of Rajasthan. Ann. Arid Zone. (2) 162 (1964).

56. C. P. Bhimaya, M. B. Jain, R. N. Kaul and B. N. Ganguli. A study of age and habitant differences in the fuel yield of Prosopis juliflora. Indian Forester 93(3) 355 (1967).

57. R. N. Kaul, R. P. Goswami, and B. K. Chitnis. Growth attributes for predicting pod and seed yield of Prosopis spicigera. Science and Culture 30(6) 282 (1964).

58. R. N. Kaul and M. B Jain. Growth attributes: their relation to fuel yield in Prosopis cineraria (Linn) McBride (P. spicigera Linn). Commonwealth For. Rev. 46(2), 155 (1967).

59. A. T. Wiley and F. G. Manwiller. Market potential of mesquite as fuel. For. Prod. J. 26(9): 48 (1976).

60. P. G. Murphy. Net primary productivity in tropical terrestrial ecosystems. In Primary Productivity of the Biosphere. H. Lieth and R. H. Whitakker (eds.) Springer-Verlag Publ. N.Y., 218 (1975).

61. W. G. McGinnies and J. F. Arnold. Relative water requirement of Arizona range plants. Technical Bull. No. 80, Univ. Arizona Agric. Exp. Sta. (1939).

62. N. I. Bazilevich and L. E. Rodin. The biological cycle of nitrogen and ash elements in plant communities of the tropical and sub-tropical zones. Forestry Abstr. 27: 357 (1966).

63. W. S. Philips. Depth of roots in soil. Ecology 44: 424 (1963).

64. O. T. Solbrig and P. D. Cantino. Reproductive adaptations in Prosopis (Leguminosae, Mimosoideae). J. Arnold Arb. 56: 185 (1975).

65. L. A. Richards (ed.). In Diagnosis and improvement of saline and alkali soils. U.S. Salinity Lab. Agric. Handbook No. 60, U.S.D.A., pp. 11 and 67 (1954).

66. R. J. Moses. In Cooling Towers. Amer. Inst. Chem. Eng. CEP Tech. Man., 42 (1972).

67. M. K. Basak and S. K. Goyal. Studies on tree legumes: Nodulation pattern and characterization of the symbiont. Annals of Arid Zone 14(4), 367 (1975).

68. D. L. Eskew and I. P. Ting. Nitrogen fixation by legumes and blue green algal-lichen crusts in a Colorado desert environment. Amer. J. Bot. (in press) (1978).

69. N. C. W. Beadle. Nitrogen economy in arid and semi-arid plant communities, III. The symbiotic nitrogen fixing organisms. Proc. Linnean Soc. of New South Wales 89, 2, 273 (1964).

70. E. R. Orchard and G. D. Darb. Fertility changes under continued wattle culture with special reference to nitrogen fixation and base status of the soil. Sixth Int. Cong. of Soil Sci. Paris IV 45, 305 (1956).

71. B. M. Sharma. Carbon-nitrogen status of soils under some plant communities of Churu, Rajasthan. Indian Forester 93(8) 552 (1967).

72. C. Charreau and P. Vidal. Influence de l'Acacia
     albida Del sur le sol, nutrition, minerale et
     rendements des mils Pennisetum au Senegal. L'Agron
     Trop. 20, 600 (1965).

73. S. A. Radwanski and G. E. Wickens. The ecology of
     Acacia albida on mantle soils in Zalingei Jebbel
     Marra Sudan. J. Appl. Ecol. 4, 569 (1969).

74. G. Jung. Influence de l'Acacia albida (Del.) sur la
     biologie des sols dior. Centre ORSTOM-Dakar, Senegal,
     West Africa mimeo, 66 pages (1967).

75. H. Jenny. In Factors of Soil Formation. McGraw-
     Hill Publ. Co. New York, N.Y., 147 (1941).

76. R. A. Date. Nitrogen: a major limitation in the
     productivity of natural communities, crops and
     pastures in the Pacific area. Soil Biol. Biochem.
     5, 5 (1973).

77. H. Breman and A. M. Cisse. Dynamics of Sahelian
     pastures in relation to drought and grazing.
     Oecologia (Berl.) 28, 301 (1977).

78. K. W. Parker and S. G. Martin. The mesquite
     problem on southern Arizona range. U.S. Dept. Agric.
     Circ. 968: 70 pages (1952).

79. H. L. Bentley. Grasses and forage plants of Central
     Texas. USDA Div. Agrostology Bulletin. 10: 36
     (1898).

80. J. G. Smith. Fodder and forage plants. USDA-Div.
     of Agrostology Bulletin 2: 56 (1900).

81. Wealth of India. Pithecellobium, Prosopis. Publi-
     cations and Information Directorate, CSIR New Delhi,
     India 8: 140 and 245 (1969).

82. J. S. Douglas. 3-D Forestry. World Crops. 19: 20
     (1967).

83. J. M. Hendershot. Ketosis in the Hawaiian Islands.
     Amer. Vet. Assn. J. 108, 74 (1946).

84. D. R. Cable. Seasonal use of soil water by mature
     velvet mesquite. J. Range Man. 30(1), 4 (1977).

85. A. R. Tiedemann and J. O. Klemmedson. Effect of mesquite trees on vegetation and soils in desert grassland. J. Range Man. 30(5), 361 (1977).

86. T. J. Peacock and C. McMillan. Ecotypic differentiation in Prosopis (mesquite). Ecology 46, No. 1 & 2, 35 (1965).

87. T. J. Peacock and C. McMillan. The photoperiodic response of American Prosopis and Acacia from a broad latitudinal distribution. Amer. J. Bot. 55(2), 153 (1968).

88. Anon. Climates of the World. U.S. Dept. of Commerce, U.S. Government Printing Office, Washington, D.C. (1969).

89. J. D. Graham. Morphological variation in mesquite (Prosopis Leguminosae) in the lowlands of Mexico. Southwestern Nat. 5(4), 187 (1960).

90. O. N. Kaul. Propagating mesquite by root and shoot cuttings. Indian Forester. 82(11), 569 (1956).

91. H. M. Hull. The effect of day and night temperature on growth, foliar wax content and cuticle development of velvet mesquite. Weeds 6, 133 (1958).

92. T. O. Flynt and H. L. Morton. A device for threshing mesquite seed. Weed Sci. 17, 302 (1969).

93. K. K. Nambiar. A novel method of improving the germination of Prosopis juliflora seeds. Indian Forester (72), 193 (1946).

94. J. M. Kingsolver, C. D. Johnson, S. R. Swier, and A. Teran. In Mesquite -- Its biology in two desert ecosystems, B. B. Simpson (ed.). Dowden, Hutchinson & Ross Inc., Stroudsburg, Pa., pp. 108 (1977).

95. The author gratefully acknowledges the encouragement, assistance, advice and use of native mesquite trees given by R. S. Bandurski, Michigan State University; G. C. Cannell, P. Copeland, D. Illes, P. Moore, J. G. Waines, P. Wilke, and D. M. Yermanos of the University of California, Riverside, and Mr. David and Mrs. Ruby Modesto of the Torres-Martinez Indian Reservation, Thermal, California, as well as the financial support of the U.S. Department of Energy Grant ET-78-G-01-3074.

6

Atriplex as a Forage Crop for Arid Lands

J.R. Goodin

Abstract

The cosmopolitan genus Atriplex consists of approximately 200 species, many of which have been recognized for
years as potential forage plants. Recent investigations
have shown that several species of saltbushes may adapt
readily to routine agronomic forage crop production. These
species are characterized by high biomass, high protein, and
mineral levels adequate for animal nutrition. Most species
are halophytes and can therefore be irrigated with saline or
brackish water, provided soil characteristics are suitable.
Such plants can expand agricultural production into areas
not heretofore considered suitable for agriculture; Atriplex
might also be used to "harvest" salt and thereby reclaim
saline land for agricultural production.

Introduction

In terms of geologic history, man's effort to domesticate plants and animals has been a relatively recent event,
but from a single generation's point of view, domestication
is a painfully slow process. There are many dead ends,
disappointments, and often social and political problems
that delay or prohibit success. Few new crops have been
overnight sensations, and the domestication process may take
hundreds of years.

With current limitations on energy resources and serious questions being raised concerning energy inputs versus
technological achievements in productivity, greater emphasis
is being given to those plants which have a potential for
productivity with minimal inputs of water, fertilizers,
mechanical harvesting and processing technology. With those
limitations, one looks to the arid and semi-arid regions of

the world for species with genetic adaptations which have assured their success over hundreds or thousands of years. Many of the desirable traits of arid species have been known for centuries, but because of ignorance or simply oversophistication in the developed countries, we have ignored many excellent possibilities.

Within the arid and saline regions of the world, the family Chenopodiaceae holds a position of distinction, and the genus Atriplex is well represented by some 200 species. Early editions of the "Transactions and Proceedings of the Royal Society of New South Wales" first reported the scientific merits of Atriplex, and by 1880 it was well known that Atriplex was a high-protein forage. Atriplex nummularia had been introduced into California by 1898 as a pasture species (1). During the early part of this century, at least five species of Atriplex were of recognized value as a forage in areas subject to summer drought for the following reasons:

(a)   the capacity for production during summer feed shortage is high

(b)   the water requirement is low, indicating a high efficiency of production relative to rainfall

(c)   the root system is deeply penetrating and capable of using moisture which has reached the subsoil during the winter

(d)   the protein and phosphoric acid contents are high

(e)   Several species produce considerably higher yields than lucerne under field conditions.

Thus, from the standpoint of mere recognition, it was apparent that Atriplex, or saltbush, exhibited characteristics which would make it desirable as an emergency feed. By 1942, Bonsma and Mare (2) were studying cactus (Opuntia spp.) and Oldman saltbush (Atriplex nummularia) in a rotational grazing scheme as alternating high carbohydrate-high protein feed sources. Since the high carbohydrate cactus was considerably more palatable than the saltbush, at least initially, the forced-feeding with saltbush created a number of difficulties with sheep. More recent studies have questioned the palatability of A. nummularia, although there can be no question about its highly digestible protein values. It has been suggested that since the tissues contain such high levels of salt, the lack of palatability may be associated with the need for a greater water intake of the animal (also a disadvantage in water-deficit areas). Milthorpe

(3) has also suggested that unpalatability might be because the proteins are too readily digested, breaking down to ammonia and being lost before they can be absorbed and causing digestive discomfort to the animals.

Taxonomy and Description

Atriplex, word of uncertain derivation, is the name of a genus of the family Chenopodiaceae which, until the time of Linnaeus included practically the entire family. Linnaeus restricted the genus to its present form. Comprised of nearly 200 species it is the largest genus in the Chenopodiaceae, a cosmopolitan family which includes over 14,000 species of herbs and shrubs.

The term saltbush has been given to this genus because nearly all of the species accumulate salt on their leaf surfaces. Some of the saltbushes are weedy herbaceous annuals, whereas others are true shrubs, some growing as tall as three meters. Saltbushes and many of their associated species belong to closely related botanical groups. Members of different species of this association often grow side by side over wide areas.

Saltbushes are closely related to the beet, chard, and spinach, and a few of the wild species are used as greens by man. They might, therefore, be expected to be good forage plants.

Observations have revealed that saltbush species often grow under conditions that will not support any other plant of economic value. It is not uncommon to find them growing in arid and often alkaline or saline soils. Many saltbush species have developed adaptations to aridity and salinity that surpass those of better known forage plants.

Chemical Analysis

The recognition that sheep, cattle, and deer have browsed various Atriplex species during the winter or in times of drought led long ago to the conclusion that the saltbushes must be nutritious, and particularly high in protein. This observation was confirmed by nutritional studies (4). Later studies on saltbushes harvested as forage (5) have confirmed these results (Table 1). Protein, total ash, calcium, and phosphorus are particularly high early in the growing season. After the first harvest, protein content declines slightly and fiber content increases considerably as the season progresses. Nitrogen free ex-

Table 1. Nutritional levels (in percent) for several
perennial forage species of Atriplex.

Species	Stage of growth	Crude Protein	Crude Fiber	Ash	Nitrogen Free Extract	Phos.
A. argentea[a]	immature	23.7	10.8	7.0	54.6	.27
A. canescens[a]	immature	14.1	16.3	10.5	57.2	.15
A. lentiformis[b]	immature first harvest	16.9	8.5	31.6	40.2	.21
	immature second harvest	14.6	17.9	23.9	42.0	.17
A. polycarpa[b]	immature first harvest	19.2	11.8	21.2	45.0	.24
	immature second harvest	15.9	20.7	17.6	43.0	.17
A. nummularia[c]	young, flow-ering shoots	21.9			41.2	
A. vesicaria[c]	leaves in drought	11.4			46.1	
A. canescens[d]	immature	19.2				

[a]From "Composition of Cereal Grains and Forages." Edited by
Committee on Feed Composition, Agricultural Board, National
Academy of Sciences, National Research Council.

[b]Goodin, J. R. and C. M. McKell. 1970. Atriplex spp. as a
potential forage crop in marginal agricultural areas. Proc.
XI Intern. Grassland Congress.

[c]Imperial Agric. Bur. Joint Pub. No. 10. "The use and misuse
of shrubs and trees as fodder." Imperial Bur. of Pastures
and Field Crops, Aberstwyth, 1947.

[d]Author's data.

tract, calcium, and phosphorus levels seem to be maintained
throughout the growing season. The evergreen, perennial
species of Atriplex maintain a source of forage during winter months in cold climates, and apparently protein levels
are sufficiently high to provide at least a partial maintenance diet.

Salt Tolerance

Many studies have indicated that plant tolerance to
chloride and sodium is associated with the relatively low
rate of ion absorption (6). The primary discriminatory
barrier to vascular transport appears to be in the roots,
and the xylem sap concentration of sodium and chloride ions
may be very low, even though accumulation may be taking
place in individual cells in the shoots (7). Plants grown
in saline media may regulate their ion uptake to a certain
extent, but generally an increase in salinity causes an
osmotic adjustment (8) and a buildup of salts in the plant
organs. The excess uptake of cations by the plant cells is
commonly associated with an increase in the synthesis of
organic acids (9, 10). Sometimes the organic acid produced
may not be favorable as far as food or forage value of the
crop is concerned. The problems associated with oxalic
acid accumulation in Atriplex will be discussed later.

In the salt economy of all plants--not only halo-
phytes--selective ion transport is of utmost importance
(11). It allows the plant to absorb from the environment
essential nutrients in amounts conducive to growth and function; it regulates the flux of ions present in the external
medium at high concentrations; and it helps to maintain
internal osmotic potentials lower than those of the sur-
roundings, and consequently, to maintain tissues in a hy-
drated condition. Even though most crop plants cannot tolerate high concentrations of salts because of specific ion
toxicities and/or osmotic effects which decrease water flow
through the soil-plant-air continuum, many halophytes can
cope with salt concentrations up to and exceeding that of
sea water (12).

Most forage species of Atriplex are considered to be
halophytes and tolerate exceptional levels of substrate
salinity (13, 14, 15, 16). Recent physiological studies
have shown that saltbushes manage to cope with high levels
of salinity by adjusting osmotically (8) and by gradually
accumulating salts in giant, vesiculated trichomes on the
surface of the leaves. If excess Na+ is accumulated, ionic
balance is achieved by synthesis of organic acids (primar-

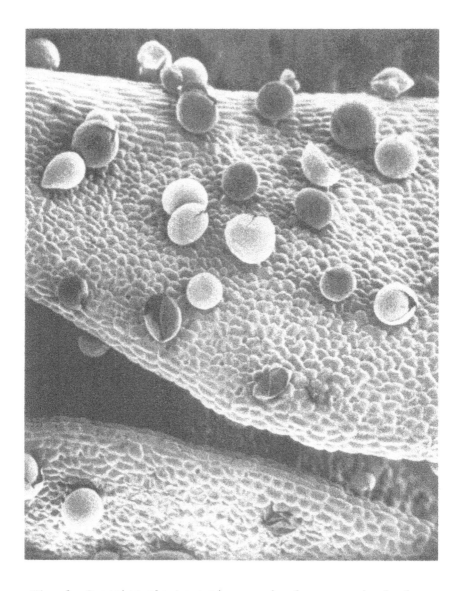

Fig. 1   Scanning electron micrograph of a young leaf of
Atriplex halimus L. showing vesiculated hairs on
the surface. 250X magnification.

Fig. 2 Scanning electron micrograph of a vesiculated hair
        on the leaf surface of <u>Atriplex</u> <u>halimus</u> L. The
        stalk cell shows the attachment to the epidermis.
        The balloon cell is approximately 800μ meters in
        diameter. 2300X magnification.

ily oxalic acid as the sodium salt) in the bladder cells
(10, 17). The concentration of salts in these cells may
reach extremely high levels, and in A. halimus values for
solute potential have actually been measured at -56,000 kPa
(15). Meanwhile, solute potentials in the remainder of the
plant tissues maintain a moderate level, allowing normal
metabolism to proceed. Eventually the vesiculated hairs
burst and leave a litter of salt crystals and cell wall
debris on the surface of the leaf. It apparently remains
there until washed off by precipitation and recycled
through the soil-plant system again (16). The conclusion
is that vesiculated hairs play an important role in the removal of salts from saltbush leaves and therefore have significance in the overall salt tolerance of the plant.

Although studied most intensely in A. halimus, apparently all species of Atriplex have vesiculated hairs on the
leaf surface and excrete salt from the plant through this
self-destructive mechanism. New vesicles develop from the
epidermis and replace the ones which have burst. Phase
microscopy and electron microscopy studies have shown that
these hair cells are alive, contain chloroplasts, and probably maintain metabolic activity by partitioning the salts
in a giant vacuole. Furthermore, the bladder cell is
attached to the epidermis by another stalk cell whose cyto-
plast contains an unusually large number of mitochondria,
indicative of high levels of metabolic activity (17, 18).

In addition to osmotic effects which lead to high
levels of substrate salinity, Atriplex manages to cope with
specific ions which are often toxic at high concentration.
Solution culture experiments with A. polycarpa (19) have
shown that species to tolerate 39,000 ppm NaCl and 80 ppm
boron. Such crops might be grown and harvested simply to
remove excess salts and thereby reclaim land too saline for
agricultural production (16).

Boyko (20) has described field experiments using saline water for crop irrigation. He concluded that sustained productivity is possible provided certain ecological and
hydrological principles are observed. Rapid percolation of
the irrigation water was essential, as was good aeration,
implying the desirability of a light sandy soil or gravel.
Boyko theorizes that the large, noncapillary air spaces in a
coarse soil provide a chamber for only partial root contact
of absorbing roots. Following percolation, the air space
becomes a humid chamber and "subterranean dew" forms on the
roots, supplying them with fresh water. This scheme would
imply that absorbing roots are constantly being subjected

to drastic changes in solute potential at the soil-root
interface. Thus, species with exceptional drought and/or
salinity tolerance would show adaptability to fluctuations
in solute potential. Under natural conditions, most sudden
changes would be in the direction of decreased salinity
caused by rainfall. The change back to the formerly high
salt content would be a relatively slow process and more or
less parallel to evaporation.

Productivity

Estimates of net primary productivity vary tremendously
with form, density, and other factors. Some plant communities have values as low as 300 kg/ha for Napiergrass
(Pennisetum purpureum) in Puerto Rico (21). A young creosote bush (Larrea divaricata) community has been estimated
at 1,000 kg/ha by Chew and Chew (22). They estimate that
the productivity will increase slightly as the number of
individuals decrease and the size of each plant increases.
In the Southwestern United States, Atriplex canescens often
occupies similar sites as creosote bush, and under similar
climatic and edaphic conditions, productivity would probably be similar. Three Atriplex species indigenous to the
western United States have been grown under routine agronomic conditions with limited irrigation, and production of
10,000 kg/ha/yr has been achieved on a sustained basis (5).
Slightly lower values have been achieved for Atriplex
canescens in Texas (23). In all of these studies, there
have been individual plants with exceptional biomass production, hence selection for high-yielding genotypes is a
definite possibility.

Establishment

Direct Seeding. Most perennial forage species of
Atriplex can be propagated by seed as well as vegetatively
by cuttings or layering (24, 25). As with other small
seeded plants, microenvironment is critical at the time of
germination. In Atriplex lentiformis, A. polycarpa, and A.
halimus, depth of planting is critical and these species
should probably never be planted more than 5-10 mm deep.
Although there is no evidence for a light requirement for
germination, the ability of the seedling to emerge from
greater depths appears to be limited. Seedling vigor is an
important part of establishment, but there is little evidence that so-called drought tolerant species have any special survival mechanism at the seedling stage. Indeed,
susceptibility to seedling mortality, whether due to desic-
cation, toxicity, insects, or diseases, is probably just as

great in xerophytes or halophytes as in mesophytes. Even
species with great salinity tolerance have difficulty coping
with salinity at the time of germination. It would appear
that establishment of such halophytes on a saline substrate
may only be possible when fresh water (e.g., rainfall)
occurs in sufficient quantity to dilute the salts during
the germination and early establishment phases. Such a
sequence of events in nature may only occur once in many
years, hence the commonly observed clustering of perennial
halophyte shrubs into similar age classes.

.Under agronomic conditions, sprinkler irrigation is
probably the easiest way, although not necessarily the least
expensive, to establish Atriplex seedlings. For most
species, seeding can best be accomplished in late fall
or early spring when evaporation rates are low. Although
overwintering may be a problem in areas where minimum temperatures fall below $-18°C$, Atriplex canescens performs
well following a fall planting provided the root system is
well established prior to dormancy.

Transplanting. Since seed germination is poor in some
perennial Atriplex species, and since seedling establishment can be a problem, greater certainty of success can be
achieved by transplanting to the field following several
months of establishment under greenhouse or lath house
conditions. We have had excellent results in planting
seeds directly into 6-cm peat pots filled with equal parts
of clay loam, peat moss, and sharp builder's sand. Germination occurs in about one week, and plants are thinned to
a single healthy seedling. These plants are grown in a
greenhouse from January until April (in Texas), and then
transplanted directly to the field at approximately 1-meter
intervals. Since there is no root disturbance, if the surrounding soil is kept moist for two or three weeks, survival rate should be near 100%. Under some conditions, it
may be necessary to harden the plants gradually by increasing light intensity and withholding water.

Weed Control

At the present time weed control is a serious problem
in Atriplex establishment. Many of the "weeds" in a field
of saltbush are close relatives and chemical weed killers
designed to control herbaceous dicots will also control the
Atriplex. Therefore, cultivation has been used to eliminate
weed competition in Atriplex research plots. In the southwestern United States, Kochia and Amaranthus species are
important weeds which compete effectively with Atriplex.

Recent evidence indicates that these species may also be important forage sources; therefore, the possibility of planting and growing two or more species at the same time should be investigated. Kochia scoparius is a particularly vigorous competitor and may out-compete Atriplex seedlings for light, water, and nutrients. In such mixed plantings, it might be possible to use Kochia as a forage in the warm growing season, and reserve the Atriplex as a high-protein winter feed.

Oxalic Acid

Many plants in the family Chenopodiaceae accumulate high concentrations of oxalic acid which may cause mammalian toxicity characterized by severe gastroenteritis with vomiting, diarrhea, melena, renal damage, and in some cases, death. The most significant accumulations apparently occur in Halogeton glomeratus, an introduced weed which accumulates as much as 30% dry weight as oxalate. There has been considerable concern that Atriplex species recom-- mended for forage can also accumulate toxic levels of oxalate. Osmond (10) has proposed that oxalate is synthesized in chenopods as the anion which balances the excess Na+ accumulated by the plant. This organic acid synthesis apparently only occurs when inorganic anions (e.g., Cl-) are somehow not accumulated at the same rate as Na+.

Recent studies with Kochia and various species of Atriplex indicate that oxalate levels are not high enough to cause toxicity problems in cattle and sheep. The assumption has been made that increasing substrate salinity leads to higher oxalate levels, but Atriplex canescens grown in culture solution accumulates less oxalate than controls whenever salinization is brought about by increasing levels of NaCl up to 0.5 M (26, 27). Ecotypes of Atriplex canescens which show great genetic variability, including selectivity for low oxalate accumulators, have been selected. There is no evidence that accumulation of either Na+ or Cl- is great enough to cause metabolic disorders.

Seed Production

Some forage species of Atriplex such as A. canescens are dioecious, and the female plants are prolific seed producers. The fruit is woody and wing-like utricles are prominent in species like the four-wing saltbush. The fruit coat and utricles contain germination inhibitors (primarily saponins) which cause dormancy and assure survival by delaying germination until rainfall has reduced the

inhibitor concentration to some safe level. Mechanical
removal of the utricles will overcome most of the inhibition, but leaching for a few hours in running tap water
will improve germination considerably. Even so, important
forage species like A. canescens germinate very poorly,
apparently due to both poor seed fill and insect damage. The
suggestion has been made that genetic problems lead to poor
fertilization and embryo development (28). The cen-
trospermous embryo is often shriveled or missing, and in
natural populations from the Southwestern United States,
germination seldom exceeds 15%. Most natural populations of
A. canescens are tetraploid, although Stutz (29) has
reported a gigas strain which is a diploid and is far more
vigorous than the common tetraploid.

Genetic studies could lead to an improved rate of fertility, and a consequent improvement in the problem of seed
sources.

There is also some evidence that insect control during
the critical stages of flowering, pollination, fertilization, and embryo development could lead to improved seed
set. We have selected some ecotypes of A. canescens which
have superior germination percentages, but it remains to be
seen whether these represent superior genetic strains or
simply years of favorable environment, insect, and disease
control.

Water Requirement

The objective in adapting Atriplex to a cropping regime is not to compete with current agronomic crops under a
high irrigation/high fertilization regime. We anticipate a
place for such halophytes in regions where minimal water is
available for irrigation, or where water quality is so poor
as to eliminate traditional crops.

Irrigation with water of reasonable quality (1200 ppm
TDS) versus brackish water (7000 ppm TDS) has shown no
decrease in yield over a period of two years (23). This
experiment was conducted on a sandy loam in a 450 mm precipitation region of west Texas. Actually, little irrigation has been needed beyond the initial establishment. A
decision to irrigate whenever the mid-day leaf water potential reached -3000 kPa turned out to be very conservative in
water consumption, and not more than 100 mm of irrigation
would be required under such conditions of environment and
edaphic factors. Increasingly saline water would require
larger quantities, and a sandier soil would also increase

the water requirement. Although reliable information is lacking concerning water use efficiency (gms water required to produce a gram of dry matter), preliminary indications are that the efficiency is very high. Even under conditions of highest productivity, there is no reason to believe that the water requirement would approach that of most mesophytic crop species. Thus, both water quantity and quality considerations place Atriplex high on the list of potential new crop plants.

Most perennial saltbushes possess C_4 metabolism which improves photosynthetic efficiency. As such, they have evolved under conditions of water stress, high light intensity, and high temperatures. Thus, one might anticipate genetic selections which are exceptional with regard to these properties.

The Crop Potential

Several species of Atriplex have been grown under routine agronomic conditions as a forage crop (5, 23). The yields are often as great as those of alfalfa, although much remains to be learned in terms of frequency of harvest, height of cut, and conditions necessary to maximize recovery.

Harvesting and transportation equipment is yet to be developed. Perhaps baling or pelleting can be utilized to reduce bulk. After a period of years, the base of the plant may become very woody unless frequency of harvest can be timed to minimize this problem. One should bear in mind that a well established plant has survival value far beyond that of traditional crop plants. Even during periods of severe water stress, the saltbush can be expected to survive.

Hundreds of reports from various parts of the world (30) classify perennial Atriplex spp. anywhere from unpalatable to highly palatable for various classes of livestock. Even the same species is reported to run the gamut of palatability. These reports must be tested in feeding trials, and the more palatable ecotypes must be selected.

Although Atriplex grows in desert soils usually deficient of nutrients, all species that we have tested respond to fertilization. There is an immediate need to study the source of nitrogen accumulated in leaf protein. Both symbiotic and non-symbiotic nitrogen fixation must be investigated to see how very large quantities of nitrogen enter the plant.

Since Atriplex is adaptable to saline and sodic soils usually not considered to be agricultural, the possibility exists that growing saltbushes on marginal agricultural land lost to production due to increasingly salty water and/or soil could be used to "harvest" salt with the crop and thereby reclaim the land for traditional agriculture. This hypothesis remains to be tested.

The demand for food from a hungry world requires that we seek every opportunity to increase production. We believe that cropping arid and semiarid regions offers one approach to that increase. By taking advantage of biological technology and innovation, we can develop new crops and new uses for old crops. Our goals must include reclamation of saline and alkali regions, and thoughtful utilization of our delicate arid ecosystems never before cropped. Through better understanding of the natural history, ecology, and physiology of the vegetation of these regions, shrubs like Atriplex may make a major contribution to productivity.

Acknowledgments

Portions of the research reported here were supported by the Institute for the Study of Physiological Stresses, Texas Tech University; Office of Water Research and Technology, U.S. Department of the Interior; and the Water Resources Center, Texas Tech University.

References

1. R. Jones, The Biology of Atriplex (CSIRO, Canberra, 1970).
2. H.C. Bonsma and G.S. Mare, Dept. of Agri. & For. Bull. No. 236, (Union of South Africa, 1942).
3. F.L. Milthorpe, The Biology of Atriplex (CSIRO, Canberra, 1970).
4. C. Wayne Cook, L.A. Stoddart and L.E. Harris, Utah State Agri. Expt. Sta. Bull. 385 (1956).
5. J.R. Goodin and C.M. McKell, Proc. XI Intern. Grassland Congress (Brisbane, 1970), p. 158.
6. H. Greenway, Aust. J. Biol. Sci. 15, 16 (1962).
7. P.F. Scholander, H.T. Hammel, E. Hemmingsen and W. Garey, Plant Physiol. 37, 722 (1962).
8. L. Bernstein, Amer. J. Botany 48, 909 (1965).
9. L. Jacobson and L. Ordin, Plant Physiol. 29, 70 (1954).
10. B. Osmond, Nature 198 (4879), 503 (1963).
11. E. Epstein, Ecological Aspects of the Mineral Nutrition of Plants (1969).
12. D.W. Rains and E. Epstein, Plant Physiol. 42, 319 (1967).
13. C.T. Gates and W. Muirhead, Aust. J. Exp. Agri. and Animal Husbandry 7, 39 (1967).
14. A. Mozafar, Doct. Diss. (University of Calif., Riverside, 1969).
15. A. Mozafar and J.R. Goodin, Plant Physiol. 45, 62 (1970).
16. J.R. Goodin and A. Mozafar, USDA Forest Service Gen. Tech. Rept. INT-1, 255 (1972).
17. B. Osmond, Aust. J. Biol. Sci. 29, 575 (1967).
18. D.B. Kelley, M.S. Thesis (Texas Tech University, Lubbock, 1974).
19. N.J. Chatterton, Ph.D. Diss. (University of Calif., Riverside, 1970).
20. H. Boyko, Salinity and Aridity (Junk Pub., The Hague, 1966).
21. R.S. Loomis and W.A. Williams, Crop Sci. 3(1), 67 (1963).
22. R.M. Chew and A.E. Chew, Ecol. Mono. 35 (4), 355 (1965).
23. D.R. Krieg, J.R. Goodin and R.G. Stevens, Rept. to USDI, Off. of Water Research and Tech. (Washington, D.C., 1977).
24. E.C. Nord, D.R. Christensen and A.P. Plummer, Ecology 50(2), 324 (1969).
25. E.C. Nord and J.R. Goodin, USDA For. Serv. Res. Note PSW 213 (1970).

26. J.R. Goodin, Proc. Symp. on Wildland Shrubs (U.S.F.S. Shrub Sciences Lab, Provo, 1975) p. 1.
27. D.K. Northington and J.R. Goodin, Proc. Symp. on Wildland Shrubs (U.S.F.S. Shrub Sciences Lab, Provo, 1975).
28. H.C. Stutz, C.L. Pope and T. Leslie, Proc. Symp. on Wildland Shrubs (U.S.F.S. Shrub Sciences Lab, Provo, 1975).
29. H.C. Stutz, J.M. Melby and G.K. Livingston, Amer. J. Bot. 62(3), 236 (1975).
30. H. Le Houerou, FAO Study No. 2 (Rome, 1969).

7

Cassia

A Tropical Essential Oil Crop

Donald L. Plucknett

Abstract

Cassia (Cinnamonum cassia Blume) is native to southern
China or northern Indochina. Less well known than its close
relative, Ceylon cinnamon (C. zeylanicum Nees), cassia none-
theless has yielded cinnamon bark and cassia oil for many
centuries. Seldom grown outside China, cassia offers promise
as a new essential oil crop in the Humid Tropics. Cassia oil
is especially prized for its rich warm flavor and fragrance,
and is used as a perfume and flavoring ingredient. Little
information is available outside China on cassia. This chapter presents information from China and elsewhere on the cultural requirements of cassia and its potential as a new crop.

Introduction

Spices have been subjects of romance and international
intrigue for many centuries, and the search for them indirect-
ly brought about great changes in international affairs, leading in part to the colonial era. Of the spices, cinnamon has
perhaps been known and sought as long or longer than any other.
To most persons the word cinnamon brings to mind the rich warm
flavor of "Ceylon cinnamon" (Cinnamonum zeylanicum Nees), and
few know of its important and less important relatives. In
particular, outside the spice trade, few persons would know
that Chinese cinnamon or cassia (Cinnamonum cassia Blume) is
one of the oldest of the spices and that much of the commercial
cinnamon bark is derived from C. cassia and not just the more
familiar C. zeylanicum.

Although in some respect C. cassia cannot be considered
a "new" crop, there have been recent developments which may
make this ancient but mysterious spice tree a potential new
crop for tropical areas where it has not been grown before.
Cassia has seldom been grown successfully outside of China,
its major area of production.

Commercial cinnamon

Cinnamon trees belong to the laurel family, Lauraceae, and there are about 50 species in the genus Cinnamonum. Most commercial cinnamon is marketed as the dried inner bark of young branches and the trunks of trees, and as was previously stated, C. zeylanicum is best known of the species. However, there are a number of other Cinnamonum species which are of importance, and the bark from some of these has frequently been confused with -- and indeed is difficult to distinguish from -- Ceylon cinnamon (1). Chief among those relatives is C. cassia (China or Chinese cassia), a native of Indochina (2) or southeastern China, its major production center. To add further to the confusion, products of other Cinnamonum species are also referred to as cassia, e.g. Saigon cassia or Saigon cinnamon (C. loureirii Nees) and Batavia cassia or Java cassia (C. burmanii Blume) (3). Also, general names for cinnamon and cinnamon products are often applied indiscriminately. The French refer to all cinnamon as "cannelle", while the Portuguese name is "canella".

C. cassia closely resembles C. obtusifolium Nees, and it has been suggested that C. cassia is a botanical variety of C. obtusifolium. Cassia is thought to be native to Indochina (7).

Cinnamon bark is best known of the several products derived from the commercial species. Cassia bark and cinnamon bark resemble each other very closely in appearance and taste, and many cinnamon-flavored products in the U.S. are flavored by cassia. In fact, the U. S. has always imported more cassia than cinnamon; in 1955 the cassia/cinnamon import ratio was 545,400 kg to 363,600 kg; in 1960, 509,100 to 272,700; and in 1972 400,000 to 263,600 (4). Of course, because of the lack of trade with China, most of the cassia imported during those years was not C. cassia; rather it was largely imported from Indonesia and a product of C. burmanii.

Cinnamon oil is an important product. It includes cinnamon bark oil, cinnamon leaf oil, and cassia oil. Cinnamon bark oil and leaf oil differ from cassia oil and from each other. In addition to cinnamic aldehyde, their major constitutuent, cinnamon bark oil and leaf oil contain eugenol, while cassia oil contains mainly cinnamic aldehyde.

Many grades and forms of bark and oil are sold commercially. The names given to these are a blend of the place of origin, the method of handling or processing, and the species used.

History—General

Cassia has been grown in China for many centuries; it was mentioned in the first Chinese herbal in 2700 B.C. (5) and was apparently used by the Egyptians in 1700 B.C. (6). Even the name cinnamon, derived probably from the Arabic "mama" or the Greek "amomum" (spice) and the prefix "chini" or Chinese, suggests that cassia was the first known cinnamon. In ancient times cinnamon was known to the Persians and Arabians as "darchini" (dar - wood or bark, and chini = China) (7).

Cassia and cinnamon are both mentioned in Sanskrit literature, the Bible in the Old Testament (8, 9) and in Greek medicinal works. The Arabs were heavily involved in the spice trade, and took pains to ensure that the origins of cinnamon and cassia were well hidden. Therefore, it was commonly thought that cinnamon came from Arabia or from somewhere else in the Middle East. In the 12th and 13th centuries A.D. cassia was common in trade in the eastern Mediterranean. The Phoenicians were also heavily involved in cassia trade. After the Portuguese found wild cinnamon in Ceylon in 1505, they forced the king to produce 113,600 kg of bark annually as tribute(5).

History in China

Cassia was known in Chinese medicine by 2700 B.C., and has long been associated with tropical and subtropical southern China. Its common name "kwei" or "kui" is given to several places, including the city of Kweilin and the Kui ling mountains in Kwangsi province (10). Bretschneider (10) prepared a comprehensive discussion of cassia in China, including a commentary on much of the ancient Chinese literature. Most of the following discussion is based on that reference.

Cassia is produced in the provinces of Kwangsi and Kwangtung, the major producing areas, and Fukien and Yunnan which produce much less (11, 12). The major production occurs near the border of Kwangtung and Kwangsi (12). In the past certain market centers (Lo Ting and Luk Po in Kwangtung; Tai On, Luk Chan and Yung Yuen in Kwangsi) were centers of cassia production (1). Lo Ting has been mentioned as the major center for cassia trade (2).

Cassia bark has been used in China for medicinal purposes, as a spice, and as a disinfectant. Bark or the other

Figure 1.   A fifteen-year-old seedling cassia tree in
            Hawaii. Mature trees become conical in shape.

products are known by various names, many of which denote
source and grade: "mou kui" ("male" cinnamon, which rolls
partly or not at all); "kui" (not rolled as a tube, but
cured inward at the sides); "kun kui" (rolled, cylindrical,
or tubular bark) -- also known as "t'ung kui" (tube); "pan
kui" (board) -- this bark does not roll and is of poor quality; "tan kui" (red bark); "mu kui" (wood) or "ta kui" (large)
-- these are taken from larger branches and are
coarse and of poor quality; "jow kui" (flesh or fleshy),
"kui chi" (branch) and "kui sin" (heart) -- these are all
barks from young twigs and are of high quality; "kwei p'i"
or "kui p'i" (cassia bark); "kui chi" (cassia twigs) and
"kui tsz" (cassia "buds", i.e. the immature fruits).

Description

Cassia is a large handsome evergreen tree which may reach
10 to 20 m in height. The tree as it matures assumes a
roughly conical shape in outline (Fig. 1). The bark is
coarse and greyish brown in color. The inner bark is brown
or red, contains the oil cells, and can be up to 13 mm thick.

Leaves are alternate or nearly opposite, stiff, thick
and leathery, oblong, near lanceolate, conspicuously 3-
nerved, 8 to 20 cm long, 4 to 5 cm wide, acuminate, basal
part cuneate, dark green and glossy above, finely hairy below, the petiole about 1 to 1.5 cm long. Flowers are in
axillary or terminal panicles about 7.5 to 15 cm long, small,
pubescent, and usually white (Fig. 2). The fruit is an oval
fleshy berry, blackish purple, one seeded, about 1 cm long,
and is borne in a shallow cup (Fig. 3). The seeds are oval,
with a conspicuously striped surface (12, 13).

The trees flower in summer (June to August) and fruits
mature in winter and spring (February to March) (12).

Cassia can be differentiated from C. zeylanicum by its
larger leaves, shorter petiole, more obtuse floral perianth,
and smaller fruits.

Habitat

Primarily a subtropical crop, cassia grows naturally in
southeastern China, Vietnam and Burma (14) and is seldom
found outside those countries, although it has been planted
expermentally in Paraguay, Sri Lanka (15) and the U.S.A.
(Florida and Hawaii). It has been cultivated on Sumatra
(13).

Figure 2.   Cassia flowers are borne in small panicles
in the leaf axils or terminally.

Figure 3.   Fruits are almost black when mature and
somewhat resemble small acorns.

The main producing areas of China are located at about 22 to 23 N latitude, while the latitudinal limit is probably about 30 N or S.

Cassia can withstand a fairly wide range of temperatures. Its native area in China experiences a wide fluctuation in temperature, both annual and diurnal. The maximum and minimum temperatures recorded during summer are 37.5 and 28 to 31 C, respectively; during winter 15.4 and -0.5 C. The monthly average maximum temperatures during the seasons are 15.4 to 18 C in winter, 23.5 to 29 C in spring, 29 to 34.5 C during summer and 26 to 29 C during fall. Monthly minimum temperatures are 10 to 12.5 C in winter, 18 to 21 C in spring, 23.5 to 26 C in summer and 18 to 21 C in fall. Average daily temperatures during the seasons are 10 to 15.4 C (winter), 21 to 26 C (spring), 26 to 29 C (summer) and 21 to 23.5 C (fall).

The crop cannot stand extremely cold weather, although it does appear that it can survive periodic light frost. The lowest temperature tolerated by cassia in China is -2.5 C (12) while in Florida it has survived -6.5 C (16). It has also survived freezing temperatures in Paraguay (18).

In China and Vietnam cassia mainly grows on hill or mountain slopes, either semiwild or cultivated.

The tree grows well where the weather is hot and humid, rainfall reaches or exceeds 1500 mm, and average annual temperatures exceed 20 C (12). Wild trees in China grow below 500 m elevation in evergreen forests. Here the weather is mainly monsoonal, and the growing season for arable crops is about 250 to 325 days.

Chinese literature frequently reports that cassia is planted mostly in moist soils in the middle slopes of low mountains, particularly on northeast slopes (12). Most planted stands are located at low elevations, about 100 to 300 m.

Soils of producing areas are those common to many tropical rain forests. They are usually oxisols or ultisols, well drained, severely leached, and of low fertility. Many are strongly acid (pH 4.5 to 5.5) (12).

The trees are quite deep rooted and therefore can withstand wind.

Older trees grow well in full sunlight, but younger trees may require shade, and indeed, are more shade tolerant.

Cassia Products

Bark

The best known products are the various barks which are used mainly in foods, confections, and medicines. Detailed descriptions of each type of bark and its manufacture are outside the scope of this chapter; however, several references exist which provide much information on cassia barks, including general descriptions and manufacturing steps (10, 17, 19); morphology, histology, and chemistry (3, 11); and grades (17, 19).

Lowest quality bark is produced at low elevations; this is the "cassia lignea" of commerce. Highest quality bark is produced at higher elevations; these are called "Kwangsi cassia".

Buds

Another well known product is "cassia buds" (1, 17) which are the dried immature fruits, including the calyx. Detailed anatomical and morphological descriptions of cassia buds are given in Parry (3).

Buds are used as spices, much like cloves, and are marketed under several names including "cassiae flores" (20), "flores cassiae" (1), "bourgeons" (17), "fleurs de cassia" (17), "kui tsz" (10), and "Kala Nagkesar" (India) (19).

Oil

Aside from bark, the product which is of most interest today is cassia oil, particularly the oil of a small group of trees belonging to a cultivar(s?) which produces a special type of oil. The principal ingredient in regular cassia oil is cinnamic aldehyde (probably the trans isomer) which may constitute as much as 90 per cent or more of the pure oil. The special cassia oil mentioned above contains the cis isomer of cinnamic aldehyde, which is used in the manufacture of a major soft drink beverage. Cis-cinnamic acid is not common in nature, the other major source being a byproduct of cocaine extraction from Erythroxylon coca (21).

In most cases cassia oil is steam distilled from the leaves and twigs, although all parts of the trees contain the oil. It is used as a flavoring ingredient and in soap and pharmaceutical manufacture.

Locally distilled oils are brownish yellow to dark brown and may contain woody resins. Pure oil is pale or dark yellow, clear, and less viscous than the crude oil. The oil has a very spicy, pungent lasting fragrance. Its taste is spicy, warm, and very sweet (1, 2, 19, 22). It is classified as a "medium-strong" flavor ingredient (22). Its suggested use level is 1 to 4 mg % and its <u>Minimum</u> <u>Percepti-</u> <u>ble</u> level is about 0.1 to 0.2 mg % (22).

Culture

Propagation

Cassia is usually propagated from seed, but sometimes vegetative propagation is also used. The trees flower in summer or early fall, and the fruits ripen in winter and spring. Fruits ripen in 2 or 3 months. In China, after picking, the outer fleshy skin is removed, washed in water and the seeds are planted immediately. Otherwise the seeds are mixed with sand and stored for a period of not more than 20 days. One kilogram contains about 2700 seeds. Germination is about 90 per cent (12).

Most seeds are germinated in special beds or pots (Fig. 4). Soil in the seedling beds should be rich, well drained, and loosely packed. The Chinese prefer to locate seedling beds on slopes where some shade can be provided by trees. The beds are usually prepared in winter. Ridges 15-18 cm high are constructed whereon seeds are planted at about 1.5 cm depth and 3 cm apart. The bed is then covered with glass to keep it moist. Seeds germinate in about 20-40 days. After germination the glass cover can be removed and a simple shade house can be constructed. Irrigation, light fertilization and careful weeding are important practices at this time.

When seedling are 15 to 18 cm tall, the shade house can be removed (remembering that some shade is still provided by trees). Seedlings are transplanted after 1 to 3 years (12, 23).

Cassia can also be propagated by cuttings. These can be single leaf, double leaf, or multiple leaf tip cuttings. Adin Steenland of Christian Mission Farms in Paraguay grows cuttings in a lathehouse, taking care to shade the cuttings from the noon and afternoon sun. He roots his plants in river sand (quartz) and uses a dilute hormone to stimulate rooting. The sand is enclosed in a wooden frame with a translucent plastic top. Watering is done hourly or as needed to keep the temperatures below 40 C, for leaf scorching occurs above 40 C. Steenland has found that winter

Figure 4.   A seedling cassia plant.

Figure 5.   A new flush on a mature tree.   Young leaves
are flaccid and soft, and when mature, become
horizontal and firm.

(August) cuttings give best results, and that 50 per cent of
the cuttings take root by Oct. 15 (spring). The Chinese
suggest that spring is the best time to take cuttings for
vegetative propagation (12).

The USDA has conducted work on vegetative propagation
(24). They point out that cassia produces flushes of growth
with 2 to 10 leaves on a branch (Fig. 5). Except for the
lower 1 or 2 leaves, the leaves are 1 to 2 cm apart at maturity, which is evidenced when leaves become horizontal and
firm. Young leaves are quite flaccid. Propagation should
wait until the flush has matured. An entire leaf flush may
be taken as single leaf cuttings, 2 leaf (2 node) cuttings,
or a single large cutting. For single leaf cuttings as much
stem as possible -- up to 5 cm -- should be used.

Cuttings must be treated with care to avoid wilting.
Mist propagation can be very successful. In cold weather
bottom heating at about 26 C may be effective.

Rooting can be slow and may require 2 to 18 months or
more to occur. Under mist propagation most cuttings remain
alive and eventually root. (In the author's experience some
leaf cuttings become diseased and leaves may become necrotic
and the cuttings die).

Rooted cuttings should be potted in a sterile medium
such as sphagnum. In potting, care should be taken that the
top axillary bud is exposed to full light, in order that it
can develop fully.

Air layering has proved successful in Hawaii (Fig. 6).
Also, in-arching can be effective.

Traditional Culture and Management in China

Cassia fruits are readily eaten by birds, and many
trees are spread and established naturally in this manner
(1). Wild trees may reach 10 to 20 m in height in ten to
thirty years. These trees are also harvested, and their bark
is highly valued for local medicinal use.

Many cassia trees are grown by families on small hillside farms (1). At least some of the trees are established
from seed, but more are probably propagated by cuttings.

Most cassia is planted in hillside fields or gardens;
often artificial terraces are constructed for this purpose.
Land preparation consists of land clearing, burning of the

Figure 6.   Young trees obtained by air layering.

debris, and preparation of the planting sites. The recommended date of planting is February 19 (the "rain water"
period) (12). Seedlings which are too tall may be trimmed.
When seedlings are used, one or two year old trees are
planted about 50 to 60 cm apart. Rows are about 1 m apart.
For essential oil (and probably bark) production the planting density should be about 9,880 trees per hectare.

Because cassia grown for medicinal purposes commands
very high prices, 5 year old trees are selected from the
denser oil or bark stands, pruned to about 1.5 m height, and
transplanted at 5 X 5 m distance (= 400 trees per ha).

Care of the young trees includes irrigation (if necessary), weeding and "adequate" fertilization (amount unspeci-
fied).

Trees grown for bark are managed under a coppice system
and are first cut when about 5 to 6 years old. At this time
the trees will be about 2 to 3 m tall. The trees are cut
off about 15 to 25 cm above the ground, and new shoots develop from the stump. The strongest three or four shoots
are allowed to grow and develop. When these shoots attain
a diameter of 2 to 4 cm or so (17), or are 3 to 4 years old
(12), they are harvested by cutting back to the stump again
and removing the bark from the shoots. This process is
continued for 50 to 60 years or until the regrowth potential
of the trees is reduced through disease or ill thrift. At
that time the fields are replanted.

Oil production is also carried out on the same coppiced
trees used for bark. Harvesting of minor branches and
leaves of trees grown for bark is done twice a year; the main
harvest being made in June and July and a smaller harvest in
January – February (1), although it has been suggested that
the best time to harvest leaves is during spring and summer
(12). The bark harvest system of cassia closely resembles
that practiced in cinnamon (C. zeylanicum) (5).

A few trees are left uncut in the field to produce seed
and cassia buds. Ford (7) reported that these trees are left
standing in regular fields at a spacing of about 15 X 30 m.
Harvesting of buds begins when the trees are about 10 years
of age. The buds are harvested in October and November when
the fruits are still developing.

Improved Management Systems for Oil Production

Work has been underway for several years in Hawaii to

Figure 7.   A fifteen-year-old tree which has been topped.
Note young flushes of leaves at the top and
upper sides.

improve cassia production for oil purposes. The concept of
the work is to develop a modified coppice management system
which is geared to oil production. In this system the trees
would be cut so as to provide maximum leaf and twig production. Trees should be spaced as close together as possible;
plant populations of 10,000 trees per ha would be desirable.
However, instead of cutting completely to the ground as in a
conventional coppice system, cutting is done to provide a
vertical framework of stems and branches from which maximum
leaf and twig production can result (Fig. 7). Thus the harvested plant should resemble a vertical "hatrack" structure
rather than a bare stump.

Because growth in Hawaii is not as seasonal as in
monsoonal China, leaf harvests should be spaced at reasonable
intervals throughout the year. Whether these intervals will
allow 2, 3 or 4 harvests which yield high quality oil has
not yet been determined.

Distillation of the Oil

Leaves, twigs, fruits and small stems can be gathered
and used in distillation. The Chinese store the leaves for
six days or more before distillation (12). This may be done
to dry the leaves, but no direct reason is given in the
literature for this practice.

Chinese references report that fresh leaves contain
about 0.3 to 0.4 % oil, the seeds 1.5% and small twigs and
stems about 1 to 2 % (12). This appears to be in conflict
with work reported by Guenther (1) in which leaves were
higher (0.54 %) in oil content than twigs (0.2 %). Mature
leaves contain more oil than young leaves. Parry (3)
analyzed various plant parts for oil content. The analyses
were as follows: bark 1.5 %, aldehyde content in oil 88.9 %;
cassia buds 0.55 %, aldehyde content in oil 80.4 %; cassia
budsticks (twigs?) 1.64 %, aldehyde content in oil 92 %;
mixture of cassia leaves, leaf-stalks and young twigs 0.77 %,
aldehyde content in oil 93 %.

In China in the past rather simple stills were set up
near streams or a water source, to provide water for steam
distillation (1). The semidry leaves and twigs were placed
in the still at a ratio of about 70 % leaves and 30 % twigs
and branches. The raw material: water weight ratio was
about 60 kg: 150 kg (1). Distillation took about 2 1/2 to
3 hours. According to Guenther (1), the oil yield from 60
kg of raw material was about 0.31 %, about half that obtained (0.77 %) from dry material in an improved still.

Overly high content of of twigs and small branches results in lower quality oil (3).

Oil quality apparently depends upon season of harvest (1, 3). Leaves harvested in midsummer and fall are reported to yield higher quality oil than those harvested in winter or early spring. This agrees with the often-reported comment that larger, more mature leaves yield best quality oil. Aldehyde content of Chinese cassia oil ranges between 70 to 95 % (1).

Much commercial cassia oil suffers from adulteration, often with kerosene and resin (1, 3). Grades of oil are usually based on cinnamic aldehyde content; for example, oils containing 80 to 85 %, to 90 %, and 90 to 95 % were differentiated in Hong Kong commerce.

Constituents of cassia oil include cinnamic aldehyde, cinnamyl acetate, phenylpropyl acetate, trace substances, salicylaldehyde, cinnamic acid (about 1 %), salicylic acid and benzoic acid, higher fatty acids, coumarin, benzaldehyde, o-methoxy benzaldehyde, and **methyl-o-coumaraldehyde**.

Potential as a New Crop

Cassia could become an important new crop in the humid tropics and subtropics, mostly as a source of desirable essential oil. In areas where labor is less expensive, bark production also may be possible. In addition, the cassia cultivar(s) which yields the cis - isomer of cinnamic aldehyde could become especially important because of its special flavoring properties.

Cassia should be an excellent crop for small farmers who cultivate upland tropical areas where oxisols and ultisols predominate, and it could become an important cash crop for such producers. In the United States cassia can be grown in Puerto Rica, Florida, Hawaii, and other Pacific island territories or possessions.

Acknowledgments

The author would like to thank Mrs. Margaret Sung White for translating major Chinese articles on cassia. Also, the help of my colleagues, Dr. Philip S. Motooka, David Saiki, Dr. Ramon S. de la Pena, Gordon Shibao and Dr. Rodolfo G. Escalada is gratefully acknowledged. Financial support came from Hawaii Agric. Exp. Sta. Project 138, Drug and Specialty Crops.

This article is published with the approval of the
Director of the Hawaii Agric. Exp. St. as Journal Series
No. 2345.

References and Notes

1. Guenther, E. The Essential Oils, Vol. IV. D. Van Nostrand
    Co., Inc., New York, p. 215-216, (1950)

2. Parry, E. J. The Chemistry of Essential Oils and Artifi-
    cial Perfumes. Vol. I. Scott Greenwood and Son,
    London. p. 549. (1921).

3. Parry, J. W. Spices; Their Morphology, Histology and
    Chemistry. Chemical Publ. Co., Inc., New York.
    p. 40-196. (1962)

4. Dull, R. E. T. Foreign Agric. p. 9-11. (May, 1974).

5. Brown, E. G. Colonial Plant Anim. Prod. 5(4):257-280.
    (1955).

6. Gildemeister, E. and F. Hoffmann. The Volatile Oils.
    Vol. I. 2nd ed. John Wiley and Sons, New York
    p. 125-129. (1913).

7. Thiselton Dyer, W. T. J. Linn. Soc. (Bot.) 20:19-24.
    (1884). The major part of this article is an account
    of a trip to cassia producing areas of China by
    Charles Ford and is sometimes cited as being authored
    by Mr. Ford.

8. Moldenke, H. N. and Alma L. Moldenke. Plants of the
    Bible. Chromia Botanica Co., Waltham, Massachusetts.
    (1952).

9. Exodus 30:23-24, Psalms 45:9, Jeremiah 6:20 Ezekial 27:19,
    Proverbs 7:7, Song of Solomon 4:14, (Jesus Sirach
    24:20-21, Apocalypse 18:13).

10. Bretschneider, E. Botanicon Sinicum. Notes on Chinese
    Botany from Native and Western Sources. Part III.
    Botanical Investigations into the Materia Medica of
    the Ancient Chinese. J. China Branch Royal Asiatic
    Soc. 24:p. 442-454. (1895). (Kraus Reprint Ltd.,
    Nendeln/Lichtenstein, 1967).

11. Nanking Herb Institute (Chinese Herbology Compiling Sec-
    tion). Zhong Cao Yao Xue (Chinese Herbology). Vol. II.
    Kiangsu People's Publisher. (in Chinese). 897 pp.
    (cassia: p. 326-330). 1976.

12. Compiling Section, Major Economic Trees. Major Econo-
    mic Trees of Southern China. Agriculture Printer,
    Peking. (1976).

13. Chun, Woon Young. Chinese Economic Trees. Commercial
    Press, Shanghai. (1921)

14. Hutchinson, J. The Genera of Flowering Plants
    (Angiospermae). Clarendon Press, Oxford. Vol. I.
    (1964).

15. Macmillan, H. F. Tropical Planting and Gardening (With
    Special Reference to Ceylon). Macmillan & Co.,
    London. 5th ed. p. 321, 382. (1956).

16. Bailey, L. H. The Standard Cyclopedia of Horticulture.
    MacMillan, New York. Vol. I. p. 773. (1947).

17. Maistre, J. Les Plantes a Epices. G.-P. Maisonneuve &
    Larose, Paris. (1964).

18. Adin Steenland, personal communication.

19. Brown, E. G. Colonial Plant Anim. Prod. 6(2):96–116.
    (1956).

20. Steinmetz, E. F. Drug Guide. Dr. E. E. Steinmetz, 347
    Keizersgracht, Amsterdam. p. XLIX. (1959).

21. The Encyclopedia Americana. Vol. 6, p. 687. (1959).

22. Arctander, S. Perfume and Flavor Materials of Natural
    Origin. Det Hoffensbergske Etalissement, Denmark.
    p. 132–133. (1960).

23. Pers. Comm., Li Chu-Shan, Tai Lung Nursery, Agric. and
    Forestry Dept., Hong Kong.

24. Pers. Comm. J. R. Haun, Leader, Chemurgic Crop In-
    vestigations, New Crops Research Branch, USDA/ARS,
    Beltsville, Maryland.

8

Guayule
Domestic Natural Rubber Rediscovered

Noel D. Vietmeyer

Abstract

Due to its superior properties, natural rubber is a
vital component of automobile, truck and aircraft tires. The
United States depends upon foreign sources of natural (hevea)
rubber for virtually all its production needs. An alternate
form of natural rubber which can be produced domestically is
derived from the guayule *(Parthenium argentatum)* plant, which
is native to Mexico and the southwestern United States.
Properties of guayule and hevea rubber are virtually identical and guayule production technology, although not fully
developed, is progressing rapidly.

Introduction

Each year the United States consumes more natural
rubber than any other nation. Virtually all tires contain
it. Automobile tires, on the average, contain 20 percent,
the new radials about 40 percent, and large tires on aircraft
and earth-moving vehicles are essentially 100 percent natural
rubber.

Virtually all of the natural rubber that we use -- half
a billion dollars' worth each year -- is imported. It comes
from the rubber tree *Hevea brasiliensis,* which can be grown
only in a few equatorial nations where rainfall is year-round
and inexpensive labor abounds. However, it is not generally
known that North America is the native habitat of an entirely
different and unexploited plant that is also rich in a natural rubber. A recent evaluation of its properties and potential suggests that it has enormous promise for large-scale
mechanized agriculture and might be economically cultivated
within our own borders.

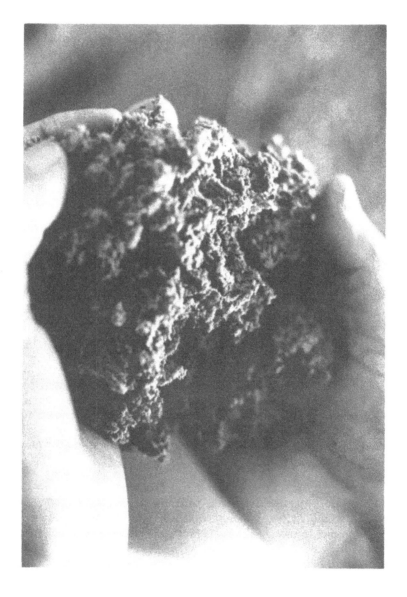

Figure 1. Guayule rubber "worms," the crude form in which the rubber separates from the wood pulp. (N.D. Vietmeyer)

The Plant

The plant is guayule *(Parthenium argentatum)* and the conclusions about its possibilities are drawn from an 18-month study of its technical qualities conducted by a distinguished committee of the National Research Council.

Guayule (pronounced wy-<u>oo</u>-lee) grows wild in wastelands of the Big Bend/Stockton Plateau region of Texas and throughout Mexico's huge Chihuahuan Desert. Indeed, in 1975 the Mexican government located in their northern states over one million hectares of guayule stands that were dense enough to be a commercially attractive rubber source.

A guayule bush, seldom more than knee-high, is a squat tuft of narrow silver-gray leaves supported on stiff gray branches. It is a true desert plant; in its native habitat rain does not fall for months, sometimes years.

When water is available, guayule grows leaves, stems, flowers, and roots, but drought or cool weather switches the plant's biochemistry to the production of rubber instead. As much as one-fourth of the weight of the stems and roots of a mature bush can be rubber. The rubber is deposited in cellular sacs, and to extract it the whole shrub must be disintegrated in the same way we pulp logs for paper. After pulping, the rubber can be floated away from the waterlogged vegetation. Alternatively, the rubber can be extracted by using solvents.

History

Guayule's history is very colorful. In 1904 John D. Rockefeller, Bernard Baruch and other New York tycoons established a company to extract rubber from wild guayule bushes. By 1910 it was the sixth largest company in Mexico, and guayule then supplied half the rubber consumed in the United States. But the wild stands were quickly depleted and Pancho Villa expelled the company from Mexico in about 1912. Subsequently, the company continued operating at Salinas, California, using shrubs cultivated from seed smuggled past Villa's alerted border guards in the tobacco pouch of William B. McCallum, the company's chief horticulturist.

For two decades the company barely survived selling small amounts of California-grown rubber. But with the fall of Southeast Asia to Japanese forces in 1942 the United States was deprived of almost its entire natural rubber supply. The federal government bought the Salinas factory

and, in a massive project involving about 1,000 researchers
and technicians as well as 10,000 workers, the Forest Service planted over 12,000 hectares of guayule in California,
mainly near Salinas and Bakersfield. However, by 1945, the
time the shrubs were maturing, the Southeast Asian rubber
plantations were back in friendly hands and hevea rubber
became amply available once more. Furthermore, the synthetic rubber industry, with massive federal price support,
was struggling to become established. A third source of
rubber therefore seemed unnecessary and the guayule fields
were burned or disked under. By late 1946 none remained.

NAS Studies

For many purposes today, synthetic rubber has proven
inferior to natural rubber. In tire carcasses, for example,
it runs hotter and degrades more rapidly than natural rubber
so that the tires develop body breaks. Also the layers on a
tire do not adhere to one another as well when made of
synthetic rubber. This is exceptionally important in the
manufacture of radial tires because the rubber is layered
onto rounded mandrils which makes delamination easier. (In
manufacturing the older tire types, rubber is layered on a
flat drum and the edges are later curved under to form the
tire's walls.) This is why the tire companies use so much
natural rubber. Radials are assuming an ever-increasing
fraction of the tire market so the future for any rubber
comparable to hevea rubber appears excellent.

It was these changed circumstances that stimulated the
National Academy of Sciences (NAS) to convene, in 1975, a
distinguished panel to review the knowledge of guayule.

Although most members of the panel had worked with some
aspect of guayule rubber production, some were invited to
serve because of their knowledge of hevea rubber, synthetic
rubber, arid land agriculture, Indian (Native American)
economic development, or developing countries. Some persons
unfamiliar with guayule -- or skeptical of its possibilities
-- were included in the panel to provide perspective and to
ensure critical evaluation of the plant's potential.

The panel's mandate was to analyze guayule's strengths
and limitations as a modern commercial crop, to identify
areas of uncertainty, and to judge the wisdom of renewed
guayule development and research. The panel was specifically
charged to consider the potentialities of guayule's providing
employment and enabling better use of land on Indian reservations in Texas, New Mexico, Arizona, and California.

The panel also gathered information from a pilot facil-
ity at Saltillo, Mexico, which is scaled to extract rubber
from 900 kg of guayule shrubs daily. In addition, just
prior to the panel's meeting, the NAS staff discovered at the
Federal Records Center near Washington, D.C., a 23-kg block
of deresinated guayule rubber produced in 1951 by the U.S.
Department of Agriculture. Samples were analyzed at Goodyear
Tire and Rubber Company, Akron, Ohio, and at Bell Telephone
Laboratories, Murray Hill, New Jersey.

The NAS study resulted in a booklet, Guayule: An
Alternative Source of Natural Rubber (1). It was the first
guayule document to be printed in more than a decade, and
the first to receive wide circulation since World War II.
Its findings have caused industrial corporations, legis-
lators, and researchers to reassess guayule's potentials.

Properties

Like hevea rubber, guayule rubber is a polymer of the
5-carbon molecule, isoprene. The isoprene units are joined
together end to end in a linear chain identical to that in
hevea rubber and with similar molecular weight. As a result,
guayule rubber has the same stretch, bounce, and general
properties as hevea rubber.

Nuclear magnetic resonance spectroscopy shows that the
microstructures of guayule rubber and of hevea rubber are
identical. The instruments detected no differences, even
though as few as five aberrant isoprene units in every 1,000
could have been detected. Essentially every isoprene unit
in guayule rubber is attached at its end and every double
bond has *cis* stereochemistry.

This contrasts with synthetic polyisoprenes, which may
contain from 1 to 8 percent of isoprene units that have aberrant connections or stereochemistry. Although this amount
appears small, it affects certain performance characteristics
beyond all proportion. This is because when rubber is
quickly stretched, any inhomogeneities destroy the ability of
the polymer strands to fit together uniformly (crystallize).

The structural similarities between hevea and purified
guayule rubber are confirmed by infrared and X-ray measurements as well as by differential thermal analyses.

Agricultural Potential

Because guayule is native to semiarid regions, it could
provide Texas, New Mexico, Arizona, and California with a new

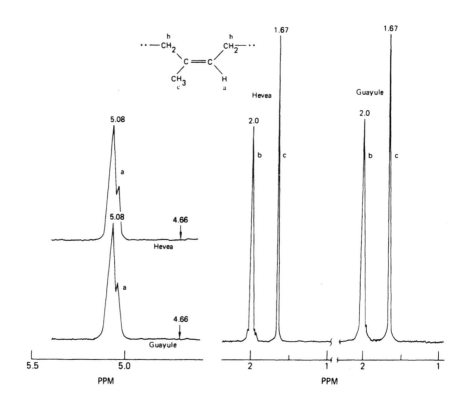

Figure 2. Carbon-13 nuclear magnetic resonance spectra confirm the structural and geometrical purity of guayule rubber,
and that guayule and hevea rubbers are, to the limits of
detection (0.5 percent), identical. (Solid samples measured
in D$_2$0 at 25 MHz, each spectrum, measured in ppm vs TMS, 4000
scans. The synthetic rubber is Li-Pi-50.) (F.A. Bovey and
E.R. Santee Jr.)

crop. The World War II project to develop guayule as an
inexpensive, domestic source of natural rubber showed that
the wild plant could be domesticated and cultivated on a
large scale. With increasing water shortages and land lying
fallow for lack of irrigation, guayule -- which probably
requires little more than the natural rainfall -- could
become a major benefit to the Southwest and to contiguous
states in Mexico. It may also grow satisfactorily in areas
that are now little used because they are too arid for conventional crops. In addition, its commercial development
could, indeed, speed the economic independence of many
Native Americans living on reservations in the Southwest.

However, no research has been carried out on guayule for
several decades so that the crop now needs an intensive
effort to apply 1978 knowledge to its cultivation and production. For example, though guayule now falls far behind
the rubber tree in yield, the two plants yielded equally well
in the 1940s and researchers see no inherent reason why
guayule cannot once again match hevea yields if given attention by plant breeders and agronomists.

Also, in the 30 years since the plant was last commercially cultivated, the methods of agriculture and industry
have become so advanced that the farming and extraction of
guayule rubber must be updated before major production can
start. For example, for each ton of rubber, the guayule
bush contains about half a ton of resin made up of various
kinds of terpenes and low-molecular-weight hydrocarbons. In
the past it was difficult to separate these resins from the
rubber, and almost all the guayule rubber sold was in a
tarry form. This gave guayule a bad reputation that still
persists in the rubber industry. It has since been found
that the resin can easily be removed by use of appropriate
solvents. But no deresinated rubber has ever been produced
for commercial use.

To improve yields and product quality, plant breeding
research is also needed.

Fortunately, the genetic diversity of the guayule plant
is substantial. Every bush growing in the desert appears
to be almost a separate strain. There is a wide variation
in rubber content (from less than 10 percent to about 26
percent) and in rate of growth, as well as in disease
resistance, ease of defoliation, ability to compete with
weeds, resin content, and cold and drought tolerance. More-
over, guayule can be hybridized with other larger species of
Parthenium. It has a unique "bimodal" reproduction system.
In one mode, the female flowers do not have to be fertilized

Figure 3. On the "O2" Ranch near Alpine, Texas, 1943. Harvesting wild guayule to provide rubber for the war effort. The arid landscape is typical of the Big Bend area where guayule is native. (U.S. Department of Agriculture)

to set seeds, and this "apomictic" reproduction ensures that once a good variety is found, it can be propagated indefinitely by seeds without genetic change. One of the greatest research needs is a simple, rapid method for screening plants for rubber content. If a suitable instrument were portable, the breeder could use it to comb wild stands for the best strains, instead of growing thousands of seeds in the blind groping for desirable types.

It is appropriate to begin the research immediately; the rubber industry projects that in the 1980s the demand for natural rubber will probably exceed the amount the rubber tree can supply. As a result, the price of natural rubber may almost double in the next few years.

Thus, with increasing rubber prices, with increasing demand for radial tires, with synthetic rubbers less competitive due to increased petroleum costs, and with huge areas of marginal semiarid land in the Southwest, a guayule comeback seems opportune.

In Mexico, guayule research based on wild plants is already well-advanced. Concerned over the poverty of compesinos in their northern desert regions, the Mexican government has financed renewed guayule development, and at Saltillo, 200 miles south of the Texas border, a dedicated team of young, highly trained scientists is applying modern methods used in manufacturing synthetic rubber. Already, more than 50 tires incorporating large proportions of guayule rubber have been made. In initial trials they have performed extremely well.

Governmental Action

The United States Congress has also seen fit to stimulate guayule's return as a crop by passing the Native Latex Commercialization Act of 1978 which makes available, through the departments of Agriculture and Commerce, $30 million in federal funds earmarked primarily for guayule research and development. This bill was first introduced in the Senate by Senator Pete Domenici (R - N. Mex.) and subsequently in the House of Representatives by Congressman George Brown (D - Cal.). It is heartening, too, that the U.S. rubber industry looks favorably on guayule. Two of the giants in the industry -- Firestone Tire and Rubber Company and Goodyear Tire and Rubber Company -- have each initiated trial plantings in the Southwest and are researching techniques for extracting rubber most economically from guayule shrubs.

References and Notes

1. Details of the guayule plant and its history, rubber extraction and processing, properties of guayule rubber, as well as selected readings, research needs and recommendations for actions to develop this crop can be found in Guayule: An Alternative Source of Natural Rubber, National Academy of Sciences, 2101 Constitution Avenue, Washington, D.C. 20418. The present review is abstracted from this report.

9

Jojoba

A New Crop for Arid Regions

LeMoyne Hogan

Abstract

Jojoba, a long-lived, dioecious evergreen shrub, is native only to certain parts of the Sonoran Desert in southern Arizona, southern California, and northern Mexico. It is the only known plant that produces a liquid wax very similar to sperm whale oil. The liquid wax is easily extracted and many uses have been documented for this unique plant product. Intensive research programs are underway at the University of Arizona, University of California-Riverside University of the Negev in Israel, and in both Sonora and Baja California, Mexico. Jojoba has great potential as a new agricultural crop for the warmer, arid regions of the world.

Introduction

In the history of agriculture, few plants have received the worldwide interest and publicity in such a brief period of time as jojoba, Simmondsia chinensis (link) Schneider.

Native to portions of the Sonoran Desert in Arizona, California, and northern Mexico (1), jojoba has captured the imagination and enthusiasm of thousands of people all over the world by its unique ability to produce fruit containing 40-60% liquid wax under relatively low moisture conditions. It remains to be seen whether this desert shrub can be profitably domesticated and if it will become a valuable new crop for arid regions of the world.

Jojoba has long been valued by native Americans and it was recognized by early explorers for its unique and valuable properties (2, 3, 4, 5, 6, 7, 8). It is evident that native stands of jojoba have been and still are an extremely valuable natural resource. They provide an almost unlimited

Fig. 1   Wild Jojoba plant near Hyder, Arizona

source of extremely variable germplasm for selection and for plant breeding. Natural stands are providing commercial quantities of seed for research and for the development of markets. It is estimated that from 10,000 to 16,000 metric tons per year are produced on wild plants in the United States and Mexico (9). Only a fraction of this has been harvested each year because of ownership, inaccessibility of stands, and the expense of wild harvesting operations. However, during the summer of 1978 the largest wild harvest in history was made.

Jojoba is a valuable browse plant for both cattle and deer (10, 11, 12), and the fruit provides a significant part of the diet of the javelina (Pecari Tajacu sonoriensis) in areas where javelina and jojoba occur together (13). Bailey's pocket mouse (Perogmathus baileyi) can live for weeks on a diet of pure jojoba seed (14).

Plant Description

The genus Simmondsia is a monotypic genus assigned to the Buxaceae family. There still remains considerable controversy as to whether or not it should be placed in a monotypic family, Simmondsiaceae, since it has few, if any, characteristics of other members of the boxwood family (1, 15, 16). Several studies have shown that 2N = 52 chromosomes in jojoba pollen mother cells. Evidence was also found which suggest that jojoba is a polypoid (17, 18, 19).

Jojoba is a many-stemmed evergreen shrub with an extremely long life. The leathery leaves vary in color from soft grey-green the first year to paler yellow-green the second year. Leaves on different plants vary considerably in size, shape, color, thickness, and pubescence. The leaves generally live through two or more seasons and continue to grow during the second season (1). In some stands old mature plants may be only 0.6 m-0.9 m high, but in more favorable areas they may grow to be more than 4.57 m high and as wide. Jojoba produces a multibranched deep-ranging root system reported to reach depths of at least 9 m (1). Although jojoba produces no well-defined growth rings and age can only be estimated (20) it is believed that plants live from 100 to 200 years.

Jojoba is dioecious; male and female flowers occur on different plants (21). The female flowers are green, urn-shaped, and rather inconspicuous. Three pistils grow from the tip of a green ovary; each ovary contains three ovules. Each ovule has the potential of becoming a seed; however, usually only one seed develops per ovary. Some plants are

Fig. 2   Female Jojoba with immature fruit

found with two- and three-seeded capsules. The male flowers
are yellow due to the large amounts of pollen released from
the pollen sacs (15). Jojoba is wind pollinated and does not
attract insect pollinators. The flowers of both sexes lack
petals, nectaries, and scent glands (15).

In Arizona, male and female jojoba flower buds form in
the axils of leaves in late summer or fall after the previous
crop has matured. They remain quiescent until the warmer,
longer days of February or March. Most pollination occurs
in March after which the pollinated flowers swell to full
size and reach maturity in July and August (15). In Israel,
under cultivation, flower buds are reported to be formed
throughout the year following peaks of vegetative activity.
This indicates that flower bud formation is largely connected
to and dependent on new vegetative growth (22), which occurs
in response to temperature and moisture. Soil moisture must
be deep as jojoba does not usually respond to light rains (1).

The most common fruiting pattern in jojoba constitutes
a single fruit which develops at alternate nodes on new
shoots. It is less common for a fruit to develop at each
node. On rare occasions, plants are found bearing double or
multiple fruit in clusters (23). At maturity, the jojoba
seeds have a tough, leathery, dark-brown seed coat. Jojoba
seeds vary greatly in size, color, and shape between individual shrubs and populations. Seeds vary in size from
700 seed/kg to 5300 seed/kg (1). Seed size reflects both the
genetic makeup of the plant and the environment. When
moisture is limiting during fruit development overall seed
size can be greatly reduced. The cotyledons are large and
make up the bulk of the seed. Cotyledon cells are filled
with liquid wax that flows freely when the seeds are cut (24)

Chemical Composition

In chemical structure, jojoba seed oil is not a fat but
a liquid wax (25). Fats, including the seed oils of all
other plants, are triglycerides (a molecule of glycerol
esterified with 3 molecules of fatty acids) (26, 27, 28).
Waxes like jojoba and sperm oil are wax esters (one molecule
of a long-chain alcohol esterified with one molecule of a
long-chain fatty acid) (26). There is a highly significant
correlation between seed weight and wax content (29). Important factors that make jojoba valuable are: (a) its purity
and molecular simplicity, (b) its stability: it can be
stored for years without becoming rancid, (c) its lubricity
after sulfurization, (d) its source of chemicals with 20 and
22 carbon atoms and, (e) its unsaturation (double bonds)(30).

Most jojoba seed are between 45 and 60 percent wax with an average of about 50 percent. Jojoba wax composition does not vary appreciably with location, soil type, rainfall or altitude (30). Wax does not change in composition as the seed matures, nor does it change during storage. The percentage of wax was found to vary inversely with the percent of gravel and soluble salts of soils of native stands, and directly with the percent of silt. Climate did not affect percentage of wax (31). Seeds analyzed 25 years after harvest show no change in wax - ester composition (30). Jojoba oil is soluble in common origin solvents such as benzene, petroleum ether, chloroform, carbon tetrachloride, and carbon di-sulfide, but it is immiscible with alcohol and acetone (32).

Jojoba can be extracted from seeds by either pressing or solvent-extraction. These two methods are used commercially to isolate vegetable oils from cottonseeds, soybeans, coconuts, and corn (32, 33).

After wax extraction, the remaining residue (meal) contains 26-32% protein as well as about 8% metabolizable carbohydrates and fiber (34). Of the essential amino acids, the lysine content is good, but the methionine content is poor (30). The nutritional effectiveness of jojoba meal in animal feeds is uncertain because of an unusual toxin in the meal. The meal contains about 4.5% simmondsin 2' furulate plus at least two other structually related compounds. The meal, when ingested by laboratory rats, causes them to avoid food, even their regular diet, to the point that they die of starvation (34). The seed hulls contain about 7% crude protein and 3% metabolizable carbohydrate (34).

Distribution

The natural populations of jojoba are found in an area comprised of approximately 38, 610 km^2 of the Sonoran desert of Mexico and the United States between latitudes 25° and 31° N and longitudes 101° and 117° west (1). Extensive native jojoba stands occur in Mexico in both northern and southern Baja California and coastal Sonora, Mexico. In California, jojoba stands occur along the southern coast as well as inland deserts. In Arizona, jojoba stands occur in southern deserts near the Colorado River eastward sporadically across most of the state and north beyond Phoenix and the San Carlos Apache Indian Reservation. Throughout the natural range of the shrub there are many separate populations varying from a few individual shrubs to as many as 500/ha.

Eleven distinct populations of jojoba have been identified in southern California. The two populations of Twenty-
nine Palms and the Aguanga, California areas are relatively
large and consist of a few thousand plants each. The remaining nine California populations are small with only a few
hundred plants in each (29).

The Twenty-nine Palms population is confined to the
Joshua Tree Monument State Park, and the Aguanga population
occurs mostly on private lands which are being rapidly sub-
divided and sold for non-agricultural purposes (29). Populations in Arizona and Mexico are much more extensive and they
constitute the largest natural jojoba areas.

Climate

Jojoba is found growing wild at elevations from sea
level on the Sonoran and Baja coast to more than 1524 m in
southern Arizona. Natural stands are believed to occur at
different elevations in response to temperature. Plants
growing at higher latitudes in Arizona are found predominately on mountain slopes with excellent cold air drainage.
They are not found in the lower valleys which receive the
cold air drainage from nearby mountain slopes. The densest
fruiting stands are usually situated on northeast, north,
and northwest-facing slopes (15). In a dry year, as in 1971-
72, plants on north-facing slopes were generally in better
vegetative condition than on other slopes, particularly those
facing south (35). At lower latitudes, as in Sonora and Baja
California with correspondingly warmer winters, they are
often found at sea level and on relatively level land.

It has been reported that mature jojoba plants will
withstand low temperatures of $-9.5°C$ and young plants will
withstand low temperatures of $-4°C$, (1), but long-term
records of temperatures found in native stands have not been
reported. Weather records have been used from the nearest
weather station which may be located several miles away and
which may represent quite different conditions.

Jojoba is generally thought to be a drought and heat
tolerant species (1). Jojoba does tolerate extreme desert
temperatures; daily summer highs of $43°C$ to $46°C$ in the shade
are common in its habitat (1). Jojoba is found in a wide
range of environmental conditions. At Twenty-nine Palms,
California the 1950-1972 average rainfall was 9 cm, the average minimum temperature was $-6°C$, and the maximum temperature
was $46°C$ (29). The other extreme is represented by the Del
Mar, California population which grows practically at sea
level with an average rainfall of 22.35 cm, an average

minimum temperature of 3.3°C, and an average maximum temperature of 35.5°C. In Sonora and Baja California it is not uncommon for jojoba to completely defoliate for several months under extreme drought conditions and to resume growth after receiving rains (1). The most productive wild jojoba is found where it receives 38 cm to 46 cm of moisture either from rainfall or runoff (1).

Soils

Jojoba is usually restricted to well-drained, coarse desert soils and coarse mixtures of gravels and clays (1). It is very tolerant of saline and alkaline soils and saline irrigation water (36, 37).

Germplasm Resources

The jojoba seed now available produces highly variable plants not only for yield but also for growth habit, fruit size, time of fruit maturity, and many other factors (38, 39). It is probable that a series of germplasm nurseries established under several environmental conditions from seed and cuttings collected from throughout the natural range of jojoba could be invaluable for selection and plant breeding (39). The first significant collection of individually selected seed was made by Gentry in 1957 (39). Ninety-two of these accessions were made available to the Plant Sciences Department at the University of Arizona in 1974 to establish germplasm nurseries. Since that time, extensive germplasm collections have been made both in Mexico and the United States (39, 40).

Natural Stand Improvement

The manipulation and management of wild populations by removing the competition of other species and by improving the moisture availability through water harvesting has been shown to improve yields in both the United States and Mexico. Erler (41) constructed microcatchments among wild jojoba plants in a marginal rainfall area (200 mm annually). Over a four-year period the average amounts of precipitation and runoff collected from October through June as a result of various soil treatments were as follows: no treatment - 154 mm; cleared the competing vegetation, smoothed and rolled - 435 mm; a water repellant coating applied - 876 mm. The seed yields in the fourth year were 27 g, 76 g, and 298 g per plant. Even if yields are increased by such treatments, it remains to be seen if manipulation and management of

wild populations will be economical. Plants grown under
these conditions will require hand picking which constitutes
a very large production cost.

Plantation Culture

It is accepted by most persons who are knowledgable
about jojoba that, if jojoba is to become an important commodity in world trade, it must be domesticated and placed
under cultivation (42). Successful establishment of plantations depends on the following: 1) The availability of
superior genetic strains for plantings, 2) the development
of suitable cultural practices, and 3) the design of low-
cost harvesting systems (9). There is strong interest
throughout the arid regions of the world in the cultivation
of jojoba. The strongest interest at present is in the
warmer areas of Arizona, California, Israel, Mexico, and
Australia (15). Test plantings have also been carried out in
Central and South America, Africa, India, and several of the
Middle Eastern countries. In the United States, several
large and small farming organizations have planted jojoba or
have plans to begin plantings in the near future (43). Many
of these groups have carefully considered the advantages and
disadvantages of jojoba and they have also studied its place
in their overall farming operation. They have the skilled
management, technology, and capital to be successful if
jojoba can be economically cultivated. There are other individuals or groups, with little or no farming experience and
insufficient capital, who are also planting jojoba on both
large and small acreages. These persons are under the impression that jojoba is a "miracle" plant which will produce
a good income with little effort; however, jojoba will
require the equivalent knowledge and good farm management as
any other perennial fruit crop presently cultivated. Jojoba
has to be treated as an orchard crop with a waiting period
for returns (44).

One who grows jojoba must be knowledgeable of its cultural requirements. He must also be extremely careful in
selecting a suitable location for planting. In addition, he
has to invest a considerable amount of work and money for
several years before the plant begins to produce (15). It
is not certain, even at that time, if there will be a profitable market for the seed.

Cold Tolerance

Jojoba is more susceptible to cold damage than commonly
understood. There are many examples where experimental and

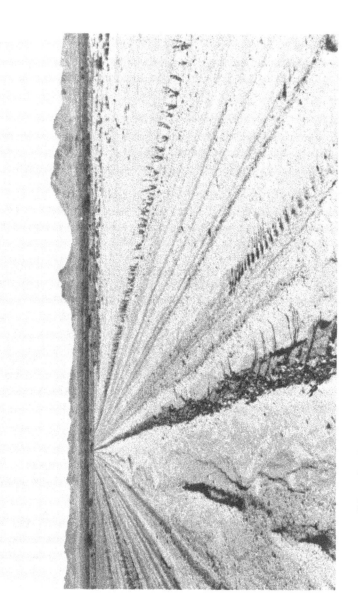

Fig. 3   One-year-old Jojoba transplants at Theba, Arizona

commercial jojoba plants have been severely damaged or completely killed in areas located near natural jojoba stands
(38).

Natural stands in Arizona and at its northern range in
California are located on mountain slopes often interspersed
with rocks. These slopes have excellent air drainage and are
usually several degrees warmer than adjacent low-lying valleys. Plantings in valleys near these slopes are often
failures as the cold air moves down the slopes and concentrates in the valleys for long periods resulting in killing
freezes. At the University of Arizona Agricultural Experiment Station at Safford all the plants in a 1-year-old germplasm nursery were killed during the winter of 1976 due to a
3-hour-period one night at temperatures of $-7.2^{\circ}C$ and a 4-
hour-period one night at temperatures of $-6^{\circ}C$. One hundred
separate accessions from 100 locations in Arizona, California
Israel, and Mexico were involved. Plants from these same
accessions, planted and handled the same at the Agricultural
Experiment Station at Mesa where the low reached $-2^{\circ}C$,
received only minimum tip damage. At the Marana plots, 90%
of the plants were killed due to low temperatures of $-5^{\circ}C$.
Yermanos (45) evaluated cold tolerance of jojoba in growth
chambers and found that 6-month-old jojoba plants subjected
to 12 hours at $-8.3^{\circ}C$ were killed. Two-year-old plants survived $-5.5^{\circ}C$ for 12 hours, but they were killed at longer
cold periods. At $-4^{\circ}C$, growing tips and young leaves of 2-
year-old plants were wilted and some older leaves developed
necrotic margins, but all plants retained vigor and continued
normal growth.

Before one chooses a commercial jojoba site one should
make certain that the area chosen is not subject to temperatures below $-4^{\circ}C$. Even if plants can be protected during the
initial stages of development, low temperatures can also
affect flowering and fruiting thus lowering the production of
nuts as well as slowing plant growth; the flowers are
damaged at $-4^{\circ}C$ to $-5^{\circ}C$ (15, 46).

Moisture Requirements

One of the most attractive features of jojoba is its
relatively low requirement for water. Positive net photo-
synthesis has been measured for naturally occurring S.
chinensis with water potentials as low as -7,000 kPa (47).
It is found growing naturally in areas where the annual rainfall is less than 12 cm per year (1). However, it grows best
and produces higher yields where it receives between 38-50 cm
of moisture per year (1). Once jojoba is well established it
can withstand extremely low moisture conditions. In Sonora

Fig. 4   Thirty-month-old Jojoba germplasm nursery at Mesa,
         Arizona

and Baja California, Mexico during extensive drought periods,
it may drop all its leaves and remain leafless for several
months only to resume active growth after a good, penetrating
rain (1). To establish a commercial planting, an adequate
supply of water must be provided especially during the first
two years. As much as 60-75 cm per year may be required to
insure that plants are not stressed (48). During the third
year, and thereafter, one can reduce the quantity of water
to 38-50 cm per year (48). Although jojoba is a deep-rooted
plant, abundant water must be available in the initial stages
to obtain a stand of seedlings (29). Established plants must
have adequate water at critical periods such as flowering and
fruit development if maximum yields are to be obtained. The
greatest need for water occurs during the late winter and
spring. It is believed that after the fruit reaches maturity
in the summer the plant benefits from a period of drought
stress. Jojoba is a drought-resistant plant, but adequate
moisture is necessary to obtain highest yields. More information is needed regarding the optimum water requirement for
jojoba under different environments, different soil types,
and for the different stages of growth. Irrigation studies
are difficult to evaluate since the variability found between
plants is often greater than the responses of those same
plants to irrigation differentials.

Fertilizer Requirements

It is difficult to measure fertilizer responses of
jojoba under field conditions with seedling-grown plant
populations. The genetic variability is very large and
genetic differences may be greater than between fertilizer
treatments.

Greenhouse sand culture studies failed to show a significant difference between six treatments of ammonium nitrate
ranging from 0 to 400 ppm during the first 124 days of growth
(49). Simmondsia chinensis had no significant response to
additions of N-P-K fertilizer in a non-irrigated field experiment. Although leaf analysis for non-fertilized plants in
the field generally showed relatively high values of N and P
in contrast to greenhouse-grown plants with high treatments
of N and P (50).

Plant Production

Jojoba is difficult to transplant (1). Any damage to
the root system of even young container-grown plants during
transplanting is often fatal. It is desirable to transplant
plants to the field when they are only a few weeks old.
Roots develop rapidly and under field conditions may

reach 60 cm before any significant top growth is evident.
This suggests that containers used for planting jojoba should
be sufficiently deep to minimize restriction of the developing roots. Unfortunately, such containers are more laborious
and more difficult to plant successfully by mechanization.
Hand planting in the United States significantly increases
planting costs (48, 51). Biodegradable open-ended containers
measuring 5 x 23 cm are adequate for starting jojoba seedlings provided the plants are transplanted within 4 months.
Planting costs can be reduced since it is not necessary to
remove such containers during transplanting. If container
grown, large plants can be successfully transplanted, although they will develop more slowly in containers than they
would under field conditions.

Unlike most oil seeds, jojoba seeds retain their viability for several years. Eleven-year-old seeds stored in an
open shed in California gave 38% germination compared to 99%
and 98% of 6-month-old seed and 2-year-old seed (1). Seeds
from 45 accessions that were 17 years old were germinated at
the University of Arizona. Germination percentages averaged
49% and ranged from 0-100% (52). Immature seed germinates
readily, but the resulting seedlings are fragile (1).

Temperature is an extremely important factor influencing
both percentage and speed of jojoba seed germination (53, 54,
55). At 26.6°C in a growth chamber, 96% of the seed tested
germinated in 10 days. If the temperature were either increased or decreased, a reduction in germination was observed
at 10 days. At 13 days, 100% of the seed germinated at
26.6°C and as temperatures either increased or decreased the
germination percentage decreased. At 32.2°C, only 58% germinated at 13 days, at 21.0°C, 83% germinated at 13 days. At
15.5°C, only 33% germinated at 13 days but at 20 days the
percentage increased to 58% (55). Therefore, it is necessary
that the soil temperature be approximately 26°C to obtain
fast germination at good percentages. This is very important
if direct seeding in the field is to be successful. Seeds
planted at lower temperatures are slower to germinate and are
often destroyed by soil microorganisms before germination
occurs.

The large amount of genetic variability found in seed-
grown jojoba indicates that some method of asexual propagation of selected high-yielding clones will be necessary if
jojoba is to reach its full potential as a cultivated crop
(1). Genetics and plant breeding can create individual high-
yielding plants, but the creation of high-yielding plants
that will reproduce true from seed may require many generations of intensive effort. The most effective, economical

way to increase yield is to select individual plants with
combinations of desirable characteristics such as yield, cold
hardiness, and upright form and to increase those plants in
commercial quantities by asexual propagation. Asexual propagation insures that growers can plant to a desired stand
with sexed plants. Possible asexual propagation methods for
jojoba are: grafting (29), stem cuttings (38, 42, 56, 57,
58), and perhaps tissue culture (59, 60). The most promising
method commercially available is propagation by stem cuttings.
Jojoba has been successfully rooted under interrupted mist
using growth regulators (56, 57, 58). Studies at the University of Arizona have shown that jojoba stem cuttings can
be propagated in commercial quantities (56). The basal portions of terminal cuttings 10 to 15 cm long of recently
matured growth are given a quick-dip of a liquid solution of
4000-8000 ppm indole-butyric acid; they are stuck, 5 cm deep,
in a mixture of 50% vermiculite and 50% perlite in a gal-
vanized metal flat each containing 100-150 cuttings. They
are placed on a greenhouse bench under interrupted mist adjusted to keep a fine layer of moisture on the foliage.
Bottom heat is provided at 23.8°C to 26.6°C. After rooting
begins a diluted solution of fertilizer is applied to promote
growth. Cuttings may root in less than 30 days, but most
cuttings require 30 to 60 days. Some cuttings have remained
under mist for as long as 12 months before rooting. For
successful rooting and establishment in soil after rooting
the foliage must remain on the cutting. If the cuttings
defoliate for any reason, they generally fail to root. During the summer of 1978 more than 70,000 cuttings from 753
wild plants from southern Arizona, selected for outstanding
potential, were made to establish a germplasm nursery. The
plants selected came from many environments and differed
greatly in appearance and condition (56). The foliage and
stems of some plants were infected to varying degrees with
several pathogenic organisms (56,61). Attempts were made
to select visibly "clean" cuttings and to disinfect cuttings
before placing them under mist. Defoliation of cuttings was
a serious problem with some stock plants. Rooting percentages ranged from 0-100% depending on the plant from which the
cuttings were made (56). Cuttings forming good roots and
retaining most of their leaves are not difficult to establish
in containers. Weak cuttings, with poorly formed root
systems, are very slow to become established and require
approximately 9 months before they can be satisfactorily
field planted.

Jojoba can be successfully V-grafted using terminal sections of 1-year-old branches (29). This procedure may possibly offer the advantage of top-working existing excess male
plants to females. The multiple stem growth pattern of

Fig. 5   Rooted Jojoba stem cuttings

jojoba reduces this methods potential for becoming commercially important. Grafting of male branches to female plants may offer a method of pollen production (29). Tissue culture may offer considerable long-term potential for increasing superior jojoba plants (69, 60, 62). While commonly used commercially for asexually increasing several florist crops and a few nursery plants (63), tissue culture with woody plants such as jojoba has had limited success. If the proper growth regulators, concentrations, and other variables can be determined and if tissue culture becomes economically feasible, uniform clones of jojoba could be reproduced quickly in very large numbers. Scientists at several institutions are seriously involved in studying this method of asexual propagation of jojoba (60, 62, 64).

It has been generally recommended that plantations should be established by nursery-grown transplants (9). If field-run wild seed is to be used for planting, there is evidence to support the desirability of direct seeding of jojoba in the field. Nursery or greenhouse production of seedlings involving the additional expenses of materials, labor, greenhouse and field transplanting costs causes direct seedings to look more inviting. If a grower has carefully selected a suitable site by virtue of soil type and climate, and if he has adequate irrigation facilities and equipment for good soil preparation, there is no reason why direct seeding cannot be superior and more profitable than nursery-grown transplants.

Selection of the planting site is of utmost importance. The soil should be well drained; it should be in an area that warms up early in the spring and doesn't go below -4°C during the coldest winters. Good seedbed preparation is as important for jojoba seed germination as it is for warm-season crops such as cotton. When the soil temperature reaches 26°C seed can be machine planted with a cotton planter by modifying the dropping plate to drop a seed each 25-30 cm at a uniform depth of 2.5 cm. Seed should be graded into at least 2 sizes and 2 plate sizes used so that a uniform stand can be obtained. The seedbed should be kept evenly moist for approximately 30 days or until a stand is established. Stored cotyledon wax is utilized by the embryo linearly during the first 30 days of germination (24). Direct seeding has been successful in Mexico, Arizona, and California where the seeding has been properly conducted at suitable sites. After the plants begin flowering, excess males and non-productive females can be rogued out leaving the desired stand. The remaining plants theoretically should be more productive than unselected seedlings transplanted to the field.

Sprinkler irrigation may prove less desirable in direct-seeded plantings as this method wets the area between the rows, and this continuous wet condition interferes with mechanical weed control and also requires more water. Either furrow or drip irrigation may be more desirable as either of these methods wets only the area adjacent to the seed thus reducing weed growth between rows.

Pruning and Training

The natural form of the majority of jojoba plants does not lend itself to mechanical harvesting. Its low-branching growth habit makes it necessary to develop some method of training jojoba into an upright form suitable for mechanical harvesting (65).

Several methods have been used (65) including staking and pruning the plant to a single leader form simulating several wild tree types found in wild stands. This procedure was soon abandoned because jojoba branches are quite brittle and a large single trunk could easily break. Lateral branching is so persistent that pruning becomes a major expense (65). It is now believed that multiple-trunked plants can be trained into an upright form by enclosing the lower few feet in a sheath of some type. Later, the lower, smaller branches can be pruned after the sheath is removed. As in many other woody plants, careful pruning of the bearing branches may be advantageous to force out new seed-bearing growth (65). This type of study can only be conclusive when genetically uniform plants are available for experimentation. Jojoba can withstand severe pruning and its growth habit can be drastically modified and adapted to meet the requirements of systems of mechanical harvesting (65).

Spacing and Pollinators

Recent trends are to grow jojoba at a final spacing between rows of 3-7 m with plants spaced 1-1.5 m on the row. The original plant spacing on the row is usually closer to allow for the removal of excess males and less-productive females by rogueing. The usual practice is to leave 1 male for each 5 females for pollination purposes (66).

Weed Control

Weed control will be necessary and it can be a significant cost factor for growing jojoba in plantations. It will be especially expensive during the first few years of growth until the jojoba plants begin to shade out the weeds. Both chemical and mechanical methods are effective. Weed control will be more difficult between rows when sprinklers are used as opposed to furrow or drip irrigation. Chemical weed control has been tested and is effective for jojoba on an experimental basis (67, 68). The chemicals and rates must be evaluated more thoroughly for different soil types and under different climatic conditions before recommendations can be made. A further complication is that no chemicals have been cleared for weed control in jojoba plantings. Clearance can be a lengthy process and depends on the acreage of jojoba requiring chemical weed control and chemical companies obtaining the required clearances for their products (68).

Diseases

Although it has commonly been stated that jojoba is relatively free of disease problems (7, 30), recent observations and studies in Arizona indicate that numerous pathogens can and do inflict serious damage under some conditions (56, 61, 69). Wild jojoba plants have not been seriously studied to determine the extent to which diseases cause death and yield reductions of native plants. As more plants have been cultivated and as scientists have observed the plant more carefully, several potentially serious jojoba diseases have been discovered (61, 69). Most of these problems have occurred either at the seedling stage or when cuttings are rooting under mist in a greenhouse. However, these same organisms also affect adult plants.

Bonar has associated *Strumella simmondsiae* with the occurrence of a jojoba leaf spot, although he did not culture the organism or confirm its pathogenicity (70). During the winter of 1976-77 at the San Carlos Apache Reservation at Bylas, Arizona, approximately 15% of 60,000 jojoba seedlings grown in containers with unsterilized soil in a shade house showed symptoms ranging from drying of leaves and defoliation to death of the plants. Symptoms were first observed on 2-month-old plants. *Phytophthora parasitica*, *Pythium aphanidermatum*, *Rhizoctonia solani*, and *Fusarium* sp. were isolated from the decayed roots (69). Twenty-nine of 72, 2- to 3-month-old jojoba plants, were innoculated with *Phymototrichum omnivorum* by burying infected sorghum seeds in the soil near the roots. They either showed symptoms of the disease or were dead after 7 weeks. Jojoba seedlings innoculated with *Verticillium dahliae* became infected as early as 3 weeks after innoculation. Severe defoliation of some jojoba stem cuttings has occurred during propagation under mist in a greenhouse. This is the greatest problem encountered in the propagation of jojoba stem cuttings (56). *Alternaria* sp. has repeatedly been isolated from leaf petioles and near nodes.

Although it has not been possible to confirm the pathogenicity by innoculations of detached leaves, the constant association of <u>Alternaria</u> sp. with defoliation suggests a cause and effect relationship (61). <u>Coniothryrium</u> sp. has been tentatively identified as being present on cuttings obtained from wild plants (61). Preplant soil treatments should be used for the nursery production of seedlings to insure against damage caused by <u>Phytophthora parasitica</u> and <u>Pythium aphanidermatum</u>; however, protecting field-established plants against <u>Phymatotrichum omnivorum</u> and <u>Verticillium dahliae</u> can be extremely difficult (61). Protecting cuttings infected with foliar disease organisms under mist can also be extremely difficult. Cuttings collected from more than 750 wild jojoba plants in the summer of 1978 varied considerably in the degree of infection depending on which plant the cuttings were taken from. The most practical method for controlling diseases at the present time is to take cuttings from those stock plants that appear to be free of foliar diseases.

Insects and Other Pests

Insects have generally not been considered to be an important problem for jojoba (9, 30). Yet as jojoba has been more closely studied and observed, a wide range of insects has been found to damage various parts of the jojoba plant. A microlepidopteron of the moth order, whose larvae chew out the young ovules and adjacent tissue in young fruits, was reported to have destroyed about 75-80% of the fruit set in Pinal County, Arizona in 1957 (71). This pest seems to be confined to elevations above 1,600 m. Recently a new interest in insects associated with wild jojoba has resulted in more extensive studies (72). These studies have identified 221 species representing 11 orders including 53 <u>Coleoptera</u> sp., 58 <u>Hymenoptera</u> sp., 24 <u>Lepidoptera</u> sp., 26 <u>Homoptera</u> sp., 19 <u>Hemiptera</u> sp., 17 <u>Diptera</u> sp., 11 <u>Thysanoptera</u> sp., 9 <u>Psocoptera</u> sp., 7 <u>Orthoptera</u> sp., 6 <u>Neuroptera</u> sp., and 1 <u>Isoptera</u> sp. Twenty-five of these species have actually been identified as feeding on jojoba. The four most important species being studied are as follows: 1) <u>Periploca</u> sp., a small leaf mining moth, 2), <u>Epinotia kasloana</u>, a moth that feeds on fruits and male flowers, 3) <u>Asphondylia</u> sp., a gall midge that causes deformation of young fruits, and 4) <u>Incisitermes</u> sp., a dry wood termite that mines stems of all sizes. During the summer of 1978 very extensive stem cutting collections were made throughout the natural jojoba stands. <u>Diaspis simmondsiae</u>, a scale, was found to be damaging wild plants as were stink bugs, cicada, and caterpillars. <u>Homalodisca liturata</u>, a leaf hopper, was found breeding in large numbers on 3-year-old nursery plants (73).

Rabbits can be a persistent problem in young plantings where jojoba may be one of the few green plants in the area. They clip off the young stems slightly above ground level often destroying the plant. It is recommended in areas of high rabbit populations that measures be taken to protect the seedlings during the first year. These measures could include using rabbit-proof fences and trapping. Ants of various species can be extremely destructive to recently planted jojoba plants. They can strip seedlings of foliage within a few hours of transplanting, thus requiring control measures in many locations. Ground squirrels will dig around new plants to get the cotyledons, destroying recently transplanted jojoba. Jojoba plantings must be fenced to keep cattle out as they relish young jojoba plants. Deer can also be a problem where other forage is limited or where deer populations are high. Since jojoba is relished by a large number of insects and animals it must be protected, especially during the first two years of growth.

Yields

Profitable commercial production of jojoba seeds requires that the average yields per given area be large enough for an economic return. A major problem facing growers is the large amount of yield variability found in plants grown from seed. Yield data from mature cultivated plants are scarce. Yield data obtained from wild plants can indicate the potential yields one can reasonably expect from cultivated plants. The weight of the seed, harvested and recorded separately from 138 wild plants in the Aguanga, California area, varied from 900 g to 5.5 kg per plant with a mean of 2.3 kg per plant (29). Wild plants averaging 2.5m in height produced 1.8 to 2.7 kg per plant (29). At the Center of Agricultural Research of the Northwest, near Hermosillo, Mexico, yields of 4-year-old cultivated plants varied from 0 to 2 kg per plant (74). First yields from the Mesa, Arizona Research plot at 2.5 years ranged from 0 to 73 g per plant. At Gilat, Israel a yield of 793 g at 4 years was obtained from the highest yielding plant and 3134 g at 9 years (22). At the Huntington Botanic Gardens at San Marino, California, 22- to 25-year-old plants have produced 15 kg (1). The Vista plots at Vista, California have provided yield data for a number of years. Four-year-old rooted cuttings of the "Vista" variety yielded 100 to 170 g each. Six-year-old "Vista" rooted cuttings averaged 350 g each. Six-year-old seedlings from Arizona seed averaged 80 g each and those from Baja, California averaged 135 g each (45). At 13 years, 13 Vista Plants exceeded 1000 g each; 3 exceeded 2000 g each, and 61 shrubs produced less than

1000 g each. Maximum yields obtained at Vista from a 6-year-old plant were 1760 g and a 12-year-old plant produced 2381 g (45).

Harvesting

For jojoba to become an economically competitive crop in the U. S., it must be machine harvested. In other countries where labor is plentiful and less expensive, hand harvesting may be economical. In Arizona, most jojoba ripens during a 3-month period, from the middle of June to the middle of September. Some plants produce seed that dehisce readily when mature; they will drop to the ground as the fingers of the harvester touch them or when the branches are shaken. Other plants produce seeds that cling tenaciously to the plant after maturity making them less easily dislodged. Therefore, selecting plants that develop mature seed uniformly with the same pattern of seed drop will facilitate the development of a mechanical harvester. Many research groups are actively developing mechanical harvesting systems for jojoba. Several mechanical harvesting systems currently used for other tree crops can be adapted to the harvest of jojoba in plantations. Plantations with well-leveled soil, slightly compacted on the surface, will be able to harvest jojoba with conventional sweeper or suction nut-picking machines such as those now used for almonds or walnuts. In areas where the soil is too sandy or powdery for such harvesting equipment, plastic netting could be used to catch falling nuts (65). It is generally believed that a satisfactory system will be available when commercial plantings are in production.

A picker can pick, on the average, 2-3 kg of mature green seed per hour in good wild stands of jojoba. Based on a wage of $2.00/hr., after drying and cleaning, the seed would have to be sold for at least .37 per kg to cover harvesting, drying and cleaning costs. The production cost of wax extracted from the seed at 40% of the clean, dry seed weight would be $1.15 per kg.

Uses

Jojoba wax shows potential in several major areas: lubrication, Factice, polishing waxes, pharmaceuticals, and cosmetics.

Sperm whale oil is widely used in lubricants because of its oiliness, its metallic wetting properties, and its non-drying characteristics that prevent gumming and tackiness. The composition and physical properties of jojoba oil closely approximate those of sperm oil. The use of jojoba oil is suggested as a substitute for sperm whale oil (30).

It has been found that sulfurized jojoba and sulfurized sperm oils are essentially equivalent in improving the load-carrying capacity under extreme-pressure conditions of both Naphthenic and bright-stock-base oils (75).

Jojoba oil can become a source of straight-chain mono-unsaturated alcohols and acids. There are no other readily available sources for these acids and alcohols. These compounds could be used as intermediates in the preparation of numerous other compounds, disinfectants, surfactants, detergents, lubricants, driers, emulsifiers, resins, plasticizers, protective coatings, fibers, corrosion inhibitors, basis for creams and ointments, antifoamers, and other products.

Long-chain unsaturated alcohols have been prepared by sodium reduction of jojoba oil (76, 77). Jojoba oil enhances the productivity of penicillin and cephalosporin antibiotics by 15-20% (78, 79). Jojoba oil reacts with sulfur chloride to form compounds with a wide range of properties, from oils to rubbery solids known as Factices. Varnishes, rubber, adhesives, and linoleum can be produced from such preparations (30, 76).

Jojoba oil can be easily hydrogenated under mild conditions using several commercial nickel catalysts. It can be hydrogenated to a hard white wax. It has properties competitive with beeswax, candelilla, carnauba, and spermaceti waxes. Hydrogenated jojoba oil has a lower melting point and is softer than carnauba. Hydrogenated jojoba oil is miscible with paraffin, solid triglycerides, and polyethylene. The main uses of saturated waxes are: floor finishes, carbon paper, and polishes for furniture, shoes, and automobiles.

A cosmetic company (Koei Perfumery, Japan) has reported the absence of any acute toxicity in laboratory and clinical tests using jojoba oil as its source in cosmetics (80).

The small quantities of wax now available and that which will come from the first cultivated plantation will be absorbed by the more lucrative markets such as cosmetics, waxes and possibly pharmaceuticals. The real challenge for jojoba will be to penetrate the larger but less lucrative market of lubricants.

Jojoba has outstanding qualities for landscape purposes in arid regions. Its value for this purpose has been studied by scientists at the University of Arizona (81, 82, 83).

The meal remaining after the oil has been extracted
should be valuable to livestock producers in arid areas
where the plant is grown. The meal contains 26-32% protein,
carbohydrates, and fiber. However, it contains simmondsin,
a toxic material which must be removed or inactivated before
it can be used as a feed (30).

Economic Potential

The major economic value of jojoba is based on the
liquid wax obtained from the mature seed. Of possible secondary importance is the meal remaining after wax extraction
as a source of animal feed. Jojoba's attraction for many
people throughout arid regions of the world is its potential
to grow and produce relatively large yields with comparatively small amounts of water. It also has the added advantage of growing and producing in saline soils with saline
water (36, 37). There are many hectares of land with these
characteristics that are now either abandoned or non-productive for crops currently grown on them.

The excellent qualities of jojoba wax have been documented (9, 30).

The future value of jojoba as a cultivated crop until
recently has been based on optimistic intuition, rather than
on firm data regarding the possibility of producing and marketing jojoba wax at competitive prices (84). Several
studies based on the plant's known requirements and the known
production costs for similar crops have provided estimated
establishment and production cost data for jojoba plantations - both on existing agricultural and raw land and for
hand harvesting and mechanical harvesting (48, 51). The high
cost of harvesting fruit from wild plants indicates the need
for jojoba to be grown in plantations and to be mechanically
harvested (84).

In the Southwestern United States the cost of buying raw
land, plus all improvements and building, or buying an established farm is about the same - $3,700 to $3,900 per ha, or
about $350,000 for 81 ha of jojoba and 8 ha for roads,
ditches, building sites, wells and other facilities (48). A
kg of jojoba oil can be produced for $4.00 by mechanical
harvesting with conservative yields of 1.36 kg of seed per
plant at maturity (12th year). If 4.54 kg of seed per plant
are harvested, this cost will drop to $2.71 per kg (48).
Seed pressed from hand-harvested seeds is considerably more
expensive, $7.45 per kg for 1.11 kg-yields and .97 for
3.73 kg-yields.

The supply of jojoba oil that will be available until about 1983 will come from seeds hand harvested from wild stands in Arizona, California, and Mexico. Plantations that are now being planted will begin producing commercial quantities of wax at that time. Clean, dry, hulled seeds were sold for extraction in 1978 for about $.74 per kg. The total cost for harvesting, cleaning, drying, extracting seed, and packaging the wax costs $2.05 to $2.23 per kg of liquid wax. The wax sold from $1.86 to $2.98 per kg because the supply was less than the demand in 1978.

The cosmetic industry is expected to be a big user of jojoba. This high-value use represents a large untapped market. Several manufacturers in the United States, Mexico, and Japan are now producing shampoo and other skin-care products (48). United States manufacturers are already marketing jojoba-based motor oils and additives (85, 86). The estimated market demand through 1981 for jojoba for both cosmetics and specialized lubricants totals 2.8 million kg annually in a price range of $1.11 to $2.23 per kg. Almost 3.7 billion kg of jojoba oil per year would be needed to replace the common oil and wax market, and 15 million kg could be sold at a price above $.37 per kg. The total penicillin industry in the United States would require 2.6 million kg of jojoba wax annually.

Summary

There is a ready market for jojoba wax. The available wax comes from hand-harvested wild plants in Arizona, California, and Mexico. In 1978 the price of liquid wax ranged from $1.86 to $2.48 per kg because the supply was less than the demand. Most of the liquid wax was used by the cosmetic industry. To capture the large-volume, low-cost lubricant market, the price of the wax must be reduced to $.55 to $.75 per kg. For this to occur, jojoba must be plantation grown and machine harvested.

Jojoba is unusual in that it is the only plant known that produces a liquid wax. It can be grown in arid regions with saline water and saline soils (36, 37).

References and Notes

1. H. S. Gentry. Econ. Bot. 12, 261. (1958).
2. E. W. Haury. The Stratigraphy and Archaelogy of Ventana Cave. Univ. of Ariz. Press, Tucson, 599 p. (1950).
3. L. J. Bean, K. S. Saubel. Temalpakh: Cahuilla Indian Knowledge and Usage of Plants. Malki Museum Press. Morango Indian Reservation, Banning, Calif. 255 p. (1972).
4. C. Niethammer. American Indian Food and Lore. Mac-Millan Pub. Co., N. Y. 191 p. (1974).
5. E. F. Castetter, R. M. Underhill. Ethnobiological Studies in the American Southwest. II. The Ethnobiology of the Papago Indians. Univ. of New Mexico Bull. Bio. Sci. 4, 84 p. (1935).
6. U. P. Hedrick Ed. Sturtevants Notes on Edible Plants. 27th N. Y. Agri. Exp. Sta. Annual Report (2). 535 p. (1919).
7. L. H. Huey. San Diego Soc. of Nat. Hist. Trans. 245 p. (1945).
8. E. K. Balls. Early Uses of California Plants. Univ. of Calif. Press. 103 p. (1962).
9. National Academy of Sciences/National Research Council, "Jojoba-Feasibility for Cultivation on Indian Reservations in the Sonoran Desert Regions." 64 p. (1977).
10. E. D. Clark. M. S. Thesis, Univ. of Arizona. (1953).
11. W. A. Dayton. Important Western Browse Plants. USDA Misc. Pub. 101, 214. (1931).
12. R. Van Dersal. Native Woody Plants of the United States their Erosion Control and Wildlife Values. USDA Misc. Pub. 303, 362. (1938).
13. T. A. Eddy. M. S. Thesis, Univ. of Arizona. (1959).
14. W. C. Sherbrooke. Ecology 57, 598. (1976)
15. W. C. Sherbrooke. Pacific Discoverer. 31, 22. (1978).
16. S. C. Brown. M. S. Thesis, Claremont Grad. School. (1976).
17. P. H. Raven, D. W. Kyhos. Aliso. 6, 105. (1965)
18. G. L. Stebbins. Chromosomal Evolution in Higher Plants. Edward Arnold (Pub.) Ltd., London. 216 p (1971).
19. R. A. Sherman, D. T. Ray. Jojoba Happenings. Univ. of Arizona 22, 2. (1978).
20. C. W. Ferguson. Jojoba Happenings 19, 16. (1977).
21. F. Gibson. Boyce-Thompson Inst., Contrib. 1, 45. (1938).
22. M. Forti. Simmondsia Studies at the Negev Institute. Nat'l. Counc. for Res. and Dev., Beer Sheva, Israel 27 p. (1973).
23. D. M. Yermanos. J. Amer. Oil Chem. Soc. 54, 547. (1977).

24. T. L. Rost, A. D. Simpson, P. Schell, S. Allen. Econ. Bot. 31, 144. (1977).
25. R. A. Green, E. O. Foster. Bot. Gaz. 94, 826. (1933).
26. N. T. Mirov. Chemurgic Digest. 9, 8. (1950).
27. D. M. Yermanos. J. Amer. Oil Chem. Soc. 52, 117. (1975).
28. T. K. Miwa. J. Amer. Oil Chem. Soc. 48, 259. (1971).
29. D. M. Yermanos. Econ. Bot. 28, 160. (1974).
30. National Academy of Sciences/National Research Council, "Products from Jojoba: A Promising New Crop for Arid Lands." 30 p. (1975).
31. W. R. Feldman. M. S. Thesis, Univ. of Arizona. (1976).
32. N. B. Knoepfler, E. J. McCourtney, L. J. Molasion, J. J. Spadaro. J. Amer. Oil Chem. Soc. 36, 644. (1959).
33. J. J. Spadaro, P. H. Eanes, E. A. Gastrock. J. Amer. Oil Chem. Soc. 37, 121. (1960).
34. A. J. Verbiscar, T. F. Banigan. In: Proceedings of the Third International Conference on Jojoba and its Uses. D. M. Yermanos, Ed. Univ. of Calif., Riverside. (1978). In Press.
35. E. F. Haase. In: Proceedings of the Second International Conference on Jojoba and its Uses. Ensenada, Mexico. (1976).
36. D. M. Yermanos, L. E. Francois, T. Tammadoni. Econ. Bot. 20, 80. (1976).
37. M. Forti, Y. Levy. Initial Response of Jojoba to Various Environmental and Cultivation Conditions. Res. and Dev. Authority, Ben-Gurion Univ. of the Negev., Beer Sheva, Israel. pp. 24. (1977).
38. L. Hogan. Crops and Soils Mag. 31, 14. (1978).
39. M. Forti. In: Jojoba and Its Uses. International Conference. E. F. Haase and W. G. McGinnies, Eds. Univ. of Arizona. 17. (1973).
40. H. S. Gentry. In: Proceedings of the Third International Conference on Jojoba and Its Uses. D. M. Yermanos, Ed. Univ. of Ca.-Riverside. (1978) In Press.
41. W. L. Ehrler, D. H. Fink. In: Proceedings of the Third International Conference on Jojoba and Its Uses. D. M. Yermanos, Ed. Univ. of Calif.-Riverside. (1978). In Press.
42. N. T. Mirov. Econ. Bot. 6, 46. (1952).
43. J. D. Johnson. In: Proceedings of the Third International Conference on Jojoba and Its Uses. D. M. Yermanos, Ed. Univ. of Calif.-Riverside. (1978). In Press.
44. L. Davidson. Rice Mill News. 11, 17. (1978).
45. E. M. Yermanos, A. Kadish, C. M. McKell, Jr. R. Goodin. Calif. Agri. 22, 3. (1968).
46. D. M. Yermanos. Mimeo Pub. Dept. Plant Sci. Univ. of Calif.-Riverside. (1978).

47. H. A. Al-Ani, B. R. Strain, H. A. Mooney. J. Ecol. 60, 41. (1972).

48. N. G. Wright. In: Proceedings of the Third International Conference on Jojoba and Its Uses. D. M. Yermanos, Ed. Univ. of Calif.-Riverside. (1978). In Press.

49. T. J. Meneley. M. S. Thesis, Univ. of Arizona. (1975).

50. J. A. Adams, F. T. Bingham, D. M. Yermanos. In: Proceedings of the Second International Conference on Jojoba and Its Uses. Ensenada, Mexico. (1976). In Press.

51. T. M. Stubblefield, N. G. Wright. Prog. Agri. in Arizona. 28, 14. (1976).

52. G. Tanner. Personal Communication. (1975).

53. D. R. McCleery. M. S. Thesis, Univ. of Arizona. (1974).

54. J. D. Burden. M. S. Thesis, Arizona State Univ. (1970).

55. T. Brown. Personal Communication. (1978).

56. L. Hogan, C. W. Lee, D. A. Palzkill, W. R. Feldman. In: Proceedings of the Third International Conference on Jojoba and Its Uses. D. M. Yermanos, Ed. Univ. of Calif.-Riverside. (1978). In Press.

57. L. Hogan, A. A. Maisari. In: Proceedings of the Second International Conference on Jojoba and Its Uses. Ensenada, Mexico. (1976). In Press.

58. A. A. Maisari. M. S. Thesis, Univ. of Arizona. (1966).

59. R. G. M. Aragao. Ph.D. Dissertation. Univ. of Arizona. (1976).

60. R. G. M. Aragao, L. Hogan. Cien. Agron. 6 (1-2), 75. (1976).

61. S. M. Alcorn, D. Young. In: Proceedings of the Third International Conference on Jojoba and Its Uses. D. M. Yermanos, Ed. Univ. of Calif.-Riverside. (1978). In Press.

62. A. Madani, C. W. Lee, L. Hogan. Hort. Science. 1, 355. (1978).

63. T. Murashige. Hort. Science. 12, 127. (1977).

64. K. J. Tautvydas. In: Proceedings of the Third International Conference on Jojoba and Its Uses. D. M. Yermanos, Ed. Univ. of Calif.-Riverside. (1978). In Press.

65. D. M. Yermanos, R. Gonzales. Calif. Agri. 30,8. (1976).

66. D. M. Yermanos. Calif. Agri. 27, 10. (1973)

67. G. S. Dhillon, D. M. Yermanos. In: Proceedings of the Second International Conference on Jojoba and Its Uses. Ensenada, Mexico. (1976). In Press.

68. R. Boyd, G. Fisher. In: Proceedings of the Third International Conference on Jojoba and Its Uses. D. M. Yermanos, Ed. Univ. of Calif.-Riverside. (1978). In Press.

69. M. E. Stanghellini. Jojoba Happenings. Univ. of Arizona. 20, 4. (1977).
70. L. Bonar, Mycologia. 34, 190. (1942).
71. R. O. Baird. Reclamation Era. July, 121. (1948).
72. J. D. Pinto, S. I. Frommer. In: Proceedings of the Third International Conference on Jojoba and Its Uses. D. M. Yermanos, Ed. Univ. of Calif.-Riverside. (1978). In Press.
73. F. G. Werner. Personal Communication. (1978).
74. M. de la Vega. Personal Communication. (1977).
75. H. Gisser. Personal Communication. (1972).
76. J. J. Spadaro, M. G. Lambau. In: Jojoba and Its Uses. International Conference. E. F. Haase and W. G. McGinnies, eds. Univ. of Arizona. 47. (1973).
77. C. Ellis. U. S. Patent. 2,054,283. Sept. 15. (1936).
78. S. G. Pathak. In: Jojoba and Its Uses. International Conference. E. F. Haase and W. G. McGinnies, Eds. Univ. of Arizona. 21 p. (1973).
79. S. G. Pathak, R. F. De Phillips, L. L. Hepler, H. E. Alburn. Proceedings of the Second International Conference on Jojoba and Its Uses. Ensenada, Mexico. (1976). In Press.
80. M. Taguchi. In: Proceedings of the Second International Conference on Jojoba and Its Uses. Ensenada, Mexico. (1976). In Press.
81. C. M. Sacamano, W. D. Jones. Univ. of Arizona College of Agri. Bull. A-82. 40 p. (1975).
82. A. E. Thompson, W. D. Jones, L. Hogan. Hort. Science. 11, 330. (1976).
83. W. D. Jones. Jojoba Happenings. Univ. of Arizona. 4. (1978).
84. D. M. Yermanos. Calif. Agri. 27, 10. (1973).
85. C. Hollingshead. U. S. Patent 3,849,323. Nov. 19. (1974).
86. C. Hollingshead. In: Proceedings of the Third International Conference on Jojoba and Its Uses. D. M. Yermanos, Ed. Univ. of Calif.-Riverside. (1978). In Press.

10

Leucaena

Versatile Tropical Tree Legume

James L. Brewbaker and E. Mark Hutton

Abstract

Leucaena is a tropical leguminous tree, Leucaena
leucocephala (Lam.) de Wit, that has a startling variety of
uses, many of them unexploited. Originating in Latin America,
leucaena was spread throughout the tropics and subtropics as
a browse legume for its protein-rich foliage. Later its
value was recognized as a nurse or shade crop for plantation
trees, and as a source of fuelwood, charcoal and roundwood.
It became widely naturalized in less acid soils of the lowland tropics, controling erosion and stabilizing forests
through its fixation of nitrogen. Its pest resistance and
durability under grazing, cutting, fire and drought have
become legendary.

Leucaena remains unexploited largely because limited
genetic variation of the genus is yet known in agroforestry.
The typical leucaenas of the tropics are seedy shrubs.
Superior "giant" varieties of leucaena have been collected
and bred in the past two decades and deployed widely. Their
wood yields are the highest reported for tropical trees.
The use of giant leucaenas is increasing as hardwood for
charcoal, high-energy fuel, plant support, fence posts, paper
pulp, chips for wood products and in reforestation and
erosion control. Other new varieties promise expanded use of
leucaena as animal forage and dried leaf meal, green manure
and fertilizer.

Introduction

History and Uses

The multifarious roles assumed by leucaena, Leucaena
leucocephala (Lam..) de Wit (1), have been the subject of
several review papers (2, 3, 4, 5, 6). They are also
Journal Series No. 2346; see note page 257.

Figure 1.   Shrubby "Hawaiian type" leucaena in its
native habitat, near Maya ruins of Labna
in the Yucatan province of Mexico.   Here
it may have served ancient Maya as fuel
wood and green manure.

discussed thoroughly in a definitive and superbly illustrated report entitled "Leucaena: Promising Forage and Tree Crop for the Tropics", published in 1977 by the National Academy of Sciences (USA) and edited by N. D. Vietmeyer (7). Leucaena (Figure 1) has been used in the past largely as a leguminous shrub for browsing animals, and as human food for its edible young seeds and pods. The forage of leucaena provides protein yields among the highest reported for plants. The tree can be grazed or cut back regularly to provide continuous high yields of a nutritious foliage that is favored by cattle, horses, pigs, sheep, rabbits and other farm animals. Dried leaf meal now commands an international market as protein and vitamin supplement for poultry and large animal feed. Studies indicate that such leaf meal could be an economic fertilizer or green manure for small farms in the tropics, and could serve as a source of liquid protein concentrates and high-protein feed for use in aquaculture.

In the future leucaena may become best known as a source of fuel or pulp wood, since its yields of hard wood can be comparable or superior to any known trees (Figure 2). Tree growth has been reported to 15 m (50 ft) in 5 years with annual wood yields in excess of 200 m^3 per hectare (50 T/A/yr bone dry weight). Since leucaena is a freely nodulating legume, fertilization with nitrogen is unnecessary. Its wood is hard and makes a high-energy fuel and charcoal. It also digests to make a satisfactory pulp for kraft paper or newspaper. The trees can be harvested as posts or poles for fences or banana props, and as lumber for parquet floors or as chips for particle board or fuel.

Leucaena is a rapidly growing, drought-tolerant tree with deep roots especially in calcareous soils. The shrubby varieties served widely as shade plants in plantations of coffee, dropping nitrogen-rich leaves to stimulate plantation yields. Such varieties also became widely naturalized especially in Pacific islands, stabilizing forests and controlling erosion, but often considered a weed to home-owners. Newer tall varieties that are less seedy are re-commended for use in reforestation and as shelterbelts and windbreaks, and can serve usefully in the role of nurse crop to forest trees like teak and eucalyptus during early growth. Leucaena is an excellent tree for reforestation of denuded, dry, nitrogen-poor tropical soils abandoned after shifting cultivation, overgrazing, or wanton deforestation and burning.

To these uses can be added the widespread custom of making jewelry from leucaena seeds, of eating the seeds and

Figure 2.   Giant arboreal leucaena variety K8
            at age 6 years in Hawaii, height 17 m.

young pods as human protein supplement, and of turning the
wood for crafts and arts.

Variation and Distribution

Leucaena (L. leucocephala) is a species native to
neutral and alkaline soils of Southern Mexico and Central
America. Other species of the genus occur as native
populations from Texas to Peru and from sea level to 2500 m
elevation. From its earliest center of diversity in Yucatan
(Figure 1) leucaena is believed to have been dispersed as a
food plant into Southwestern Mexico. Young seeds are eaten
raw or cooked, and pressed into beancakes today in states
such as Guerrero, Chiapas and Oaxaca. The state of Oaxaca
derives its name from leucaena or "huaxin" (Maya). In the
Spanish galleon trade leucaena came to the Philippines
probably before 1600, evidently as feed and bedding for
horses accompanying Spanish gold to pay for goods obtained
from the Orient. It thence spread throughout the Pacific
and Southeast Asia.

In the 19th century leucaena was dispersed further as a
shade or nitrogen-nurse tree for plantation crops like
coffee, cacao, quinine and pepper, notably in Indonesia and
Africa. Now it is naturalized throughout the tropics
between 30°N and 30°S latitudes on less acid soils at
elevations up to 1500 m.

Most leucaenas in the world are very similar to the
freely-seeding, shrubby type of the west coast of Mexico.
This type is often called "Hawaiian", in reference to its
domination of lowland forests in Hawaii. Leucaena is a
self-pollinating plant, and many populations in the world
probably trace back to trees from the region of Acapulco,
Mexico, major port for the Spanish galleon trade.   Hawaiian
and Australian investigators have been assembling leucaena
germplasm since 1960, with the first extensive field
collection made in 1967 under auspices of the USDA by the
senior author, and continued with public and private funds in
1977 and 1978*.

Genetic variation of Leucaena leucocephala in the field
is complemented by that of other species in the genus, to be
described, which extend the range of leucaena to cooler

------

*The senior author acknowledges especially the support
of the Germplasm Resources Unit of the U.S. Dept. Agric., and
of the General Service Foundation, John Musser, President,
for these explorations.

montane climates and perhaps also to the acid, aluminous,
and waterlogged soils of the tropics.

Current Status

Leucaena is disappearing rapidly from its center of
origin, like many other leguminous trees of the tropics.
Uncontrolled deforestation, burning and grazing have
eliminated many native populations from Mesoamerica in
recent years. Naturalized populations occur throughout the
world, but may contribute little new genetic variation, if
as suspected they trace back to a limited genetic base.
Thus a world collection and gremplasm bank of leucaenas has
been established in Hawaii, under the University of Hawaii
and the USDA.

Leucaena first entered world commerce in 1974 with the
production and export of dried leaf meal for animal feed
from the Philippines and Malawi. Large experimental plantings of leucaena as a pasture legume have been made in
Mexico and Australia. A few large plantings were underway
in the Philippines by 1979 to provide charcoal as fuel in
manufacture of plastics in the Philippines and of steel in
Japan. Wood and pulp properties of leucaena are undergoing
tests by industry, and limited silvicultural trials are
underway to determine optimal management practices for high
wood yields.

In perhaps its most important role as a crop for the
small tropical farmer, however, leucaena receives little
attention. As a source of fuel wood and green manure and
animal feed on the small farm, leucaena is unexcelled in much
of the tropics. Increasing tropical deforestation promises
to make many farmers expend as much energy harvesting fuel
and wood products as in food crop management. Leucaena's
agroforestry potential for the small farmer in the tropics
thus has been addressed in detail in this paper.

Botanical Description

Species Variation

The common name "leucaena"* is applied here to one
species, <u>Leucaena leucocephala</u> (Lam.) de Wit. The species
has been known under several other names, including <u>L</u>. <u>glauca</u>
(Willd.) Benth, <u>L</u>. <u>latisiliqua</u> (L.) W. T. Gillis, <u>L</u>.

---

*Correctly pronounced "loo-see-nah", but also pronounced
"loo-say-na" and "loo-key-nah".

salvadorensis Standl. and L. glabrata Rose (1). Many common
names (6) are applied to leucaena around the world, notably
guaje, huaxin and uaxim (Mexico and Central America), lead-
tree, tan-tan, hediondilla and white popinac (Caribbeans),
ipil-ipil and bayani (Philippines), lamtoro and lanang
(Indonesia), koa haole (Hawai   vaivai (Fiji) and tang-a-tang
(Guam).

Leucaena is a member of the Linnaean genus by the same
name, and represents the tribe Mimoseae of the sub-family
Mimosoideae of the family Leguminosae*. Our field studies
confirm the status of nine other species in this genus
(Table 1). Several of these sepcies deserve agroforestry
evaluation, notably L. esculenta, L. macrophylla, L. trichodes,
L. pulverulenta, and L. diversifolia, whose natural variation
is probably the greatest in the genus. An additional 42
species have been attributed to this genus, but are
presently viewed as synonymies of those in Table 1 (8).

The Mimosoideae comprises tropical legumes marked by
their arboreal habit, regular flowers in round heads, finely
divided leaves, and pods that are conspicuously flattened
and often curled. The Mimosoideae has attracted very
limited research by agriculturist or botanist. Species of
Acacia, Albizia, Enterolobium, Inga, Leucaena, Pithecellobium,
Prosopis, Samanea and Xylia have been deployed to some extent
for lumber, shade, food uses and in reforestation. Mimosoid
genera of importance for future studies include Calliandru,
Entada, Mimosa, Parkia and Piptadenia, most of them American
genera. The chromosome numbers of mimosoids have a common
base of 13, distinguishing them from most true legumes
(Sub-family Papilionatae), with a base number of 7 or 8.

Genetic Variation

The most striking genetic variation within the Leucaena
species is that of plant and leaf size and habit. Tall
arboreal variants and dwarfed, shrubby forms appear in most
species. These have been designated the "Salvador type" and
the "Hawaiian type" in leucaena (Figure 3). Nearly all
collections of leucaena made both within and outside Latin
America are of the Hawaiian type, small trees to 6 m in
height that normally start flowering within 6 months of
planting. The Hawaiian type is a small, highly branched
version of the Salvador type in most quantitative traits,
with small leaves and leaflets, small pods and seeds,
narrow trunks and branches (Figures 1, 2, 3).

---

*Alternatively, systematists place it in the sub-family
Mimosoideae of the family Mimosaceae.

Table 1.   The ten major species of Leucaena.

Leucaena Species	Date of Epithet	Locale	Elev.	Tree Ht. -m-	Forestry Potential
leucocephala	1763	Mex./C. Am.	Low	22	Exc.
esculenta	1875	Mexico	High	25	Exc.
trichodes	1842	So. Am.	Low	22	Good
macrophylla	1844	W. Mexico	Low	18	Good
Collinsii	1928	So. Mexico	Mid	13	Good
diversifolia	1842	Mex./C. Am.	High	18	Good
pulverulenta	1842	No. Mexico	Low	20	Fair
Shannoni	1914	C. Am.	Low	9	Poor
lanceolata	1886	E. Mexico	Low	9	Poor
retusa	1852	Texas	Low	8	Poor

Table 2.   Varieties of leucaena tested widely since 1970.

Variety	Origin	Type	Comments
Peru	Argentina	Peru	Good forage type
Cunningham	Australia	Peru	High-yielding forage type
K8	Mexico	Salvador	Uniform, low-seedy "giant"
K28	Salvador	Salvador	Similar to K8
K29	Honduras	Salvador	Very low seediness and limbiness
K67	Salvador	Salvador	More genetic variability
K72	Hawaii	Salvador	More seediness
K132	Mexico	Salvador	Very long pods

Figure 3. Pods, seeds and leaves of "Hawaiian type" leucaena
(3A), and giant or "Salvador type" (3B).

Figure 4.   Species hybrid trees of variety K340, Leucaena
pulverulenta X L. leucocephala, at age of 3.5 years.

The Salvador type was first recognized by botanists as a distinct species under the name "L. glabrata", and later by foresters in Indonesia where it is called "lanang" (3). It no longer appears to exist in native populations, although regions of southern Mexico and western Central America may yield such populations. Salvador-type trees grow to 20 m, averaging more than 4 m per year under suitable conditions. These giant types have been dubbed "Hawaiian Giants" in Hawaii (9), and seeds of outstanding varieties K8 and K132 (Mexico), K28, K67 and K72 (Salvador), K29 (Honduras) and K62 (Ivory Coast) have been dispersed since 1968 in the tropics for further evaluations (Table 2).

A third type of leucaena, designated the "Peru type" (10), combines the larger leaves and vegetative vigor of the Salvador type with extensive low branching and reduced relative woodiness. Varieties of this type like "Cunningham" have been bred in Australia and appear excellent for forage harvest and grazing (11). Although referred to as the Peru type, this form of leucaena traces to seeds from Argentina of suspected Peruvian origin (12), but the morphotype appears to occur naturally only in Nicaragua. It may be a product of hybridization of the Salvador and Hawaiian types.

Leucaena (L. leucocephala) is highly self-fertilized and produces seeds abundantly, and can thus be propagated as true lines from seed. Breeding cycles are rapid, between 10 and 15 months from seed to seed. Varying amounts of cross-fertilization evidently occur in other Leucaena species. Most crosses among the ten species appear fertile (13). Crosses of L. pulverulenta (2n=56) with L. leucocephala (2n=104) are often attractive and vigorous trees with potential silvicultural value (Figure 4). Advanced generations of this cross may also produce varieties of value as forage. Chromosome counts of 4 other Leucaena species were uniformly 2n=52, suggesting leucaena to be polyploid within the genus. Cytological and genetic studies, however, suggest disomic pairing (13, 14).

Gray (12, 14) studied the inheritance of several traits in leucaena. The Salvador-type arboreal habit was shown to be dominant over the Hawaiian or shrub type, and controlled by one major gene in crosses studied. Similarly the low-branching habit of the Peru type appeared to be recessive to the Salvador arboreal habit, while progenies from Peru X Hawaiian crosses showed no simple genetic ratios. Quantitative variations in plant vigor and time of flowering were inherited polygenically. Genetic studies on leucaena were also reported in Hawaii (13), and polygenic control implied for variations in protein and mimosine contents. Diallel

Figure 5. Growth of leucaena variety K8 following
           transplanting of 3-month old seedlings
           in Hawaii. 5A age one month; 5B age
           3 months; 5C age 5 months.

studies of stem length and stem number (14) indicated high
general combining abilities in four varieties of contrasting
growth habit.

Growth Habit

Leucaenas have diverse growth habits that adjust to a
remarkable variety of environments and management systems.
Under careful management the Salvador-type tree grows at a
rapid pace from transplanting to mature height, attaining 1 m
in one month, 3 m in 3 months and 5 m in 5 months (Figure 5).
Isolated trees grew in 3 years to 12 m in height (Figure 4),
and in 6 years to about 20 m in height and 25 cm in diameter,
with crown spreading to 15 m. When cut to the ground, the
tree can produce a cluster of branches to 10 m in length
within 2 years. If planted in a dense stand and regularly
cut to the ground or grazed, it can be maintained for decades
as a low, foliose shrub. Natural populations of the Hawaiian
type tend to overpopulate through reseeding and produce
thick stands of spindly plants, since light readily penetrates
through the canopy. Salvador-type plants grow tall with more
dense overstory, and overpopulation appears less of a
problem with such varieties.

The leucaena leaf and flower are distinctively
Mimosaceous (Figure 3). The leaves are bipinnately divided
into very fine leaflets 8 to 15 mm in length, and provided
with 11-22 pairs of leaflets per branch (pinna), and 4-9
pairs of pinnae per leaf. The leaves bear distinctive
bathtub-shaped glands near the base of the petiole, and the
leaflets sleep or fold in pairs at night. Trees remain
evergreen when moisture is not limiting, with individual
leaves surviving between 2 and 6 months.

The white regular flowers of leucaena are packed up to
180 in dense globular heads to 2.5 cm in diameter when pollen
are being shed. Other L. species have white, yellow or
reddish heads of varying sizes and seasons of appearance.
The flowers are somewhat fragrant. Hawaiian-type leucaenas
flower heavily through the year when moisture is not limiting,
but Salvador-type trees tend to flower seasonally.

Pods of leucaena are distinctively flat, brown, hanging
straight down, with parallel margins that separate when ripe.
Pods and seeds of Salvador types are much larger than those
of the Hawaiian for the two types (Figure 3). The elliptic
brown seeds are arranged transversely in the pods, and bear
an impervious seedcoat that must be scarified to provide
uniform germination, as described later. The raw young seeds,
that are a favored snack food in Southern Mexico, contain over

30% protein. Dry seeds include about 50% seedcoat.

The young seedlings root aggressively to produce a single tap root with a few sharply-angled lateral roots. Roots penetrate friable soils to a depth of a meter within two months of germination. The roots are deep and never or rarely superficial, making it a good windbreak and companion tree. Leucaena roots nodulate in the presence of the right rhizobia, and inoculation occurs naturally in many tropical soils. The nodules are confined to upper layers of soils, but can account for fixation of more than 500 kg/ha per year of nitrogen when trees are cut regularly as forage. Leucaena rhizobia are typically acid producing, but strain CB81 from Australia functions better on acid soils (15).

Leucaena produces a dense, yellowish hardwood more fully described in a later section. The trees grow back or coppice rapidly following cutting, allowing repeated harvests for firewood and other uses. Cuttings do not root readily, and improved methods are needed to facilitate vegetative propagation. Bud grafting can be done with evident ease (3).

Environmental Requirements

Limiting Factors

Factors which have limited the distribution and use of leucaena are political and sociological as well as ecological and biological. Fuel crops of the tropics are notoriously understudied, and the growing of fuel has not been considered necessary or desirable by most tropical cultures. This cannot wisely be allowed to continue. The new "giant" types of leucaena have been distributed widely only since 1968, and are unknown in many tropical countries. The use of a high-protein forage crop like leucaena is also undervalued in most of the tropics, where animal feeds give necessary precedence to food crops, and where understanding is often lacking of the significance of quality as opposed to quantity in a forage crop. The mild toxicity of leucaena, due to its alkaloid mimosine, further discourages its use. Seediness of common leucaenas and their tendency to become weedy poses a limitation that does not appear to apply to the "giants", but delays their acceptance. Leucaena is an unknown tree to most of the big forestry industries of the world, and there is natural reluctance to change from pines, eucalypts and other favorites to a relatively unstudied new tree, despite apparent advantages.

There are several major biological and ecological
constraints on the spread of the leucaenas. These include
the soil and temperature restrictions discussed in the
following section. Pests and diseases of the leucaenas have
been very limited, no doubt a reflection in part of the lack
of relevant research on them. Only one disease is known to
cause yield loss in the field, the leucaena leafspot
(Camptomeris leucaenae). Leafspot is sporadic in the
Caribbean, Yucatan peninsula, and in northern S. America, and
research is needed on its biology, potential severity, control
and resistance. The seeds of leucaena are invaded by seed
weevils throughout the world, notably the species Pseudococcus
citri and Ferrisia virgata. These pose a problem for seed
producers, and may cause damage on crops nearby (3). The
seed weevils are rarely common when pods are newly ripened,
becoming severe as pods are allowed to linger on the trees.
Wood borers have been reported rarely on the leucaenas;
the black twig borer, Xylosandrus compactus, causes sporadic
severe damage in Hawaii.

Climate and Soil

Leucaena is strongly perennial and continues its growth
throughout the year, but rapid dry matter production is
dependent on a favorable combination of temperature and soil
moisture. Thus under hot dry conditions, or where there is
a succession of frosts, all the leaves are shed and growth
almost ceases. In the wet tropics constant prolific growth
is possible on fertile soils or where soil nutrient deficiencies have been corrected on poor soils. The optimum
temperatures for leucaena growth appear to be between 25° and
30°C, with very little growth occurring below 10°C. It will
survive frosts provided they are not too heavy and frequent.
For example the woody above-ground framework can be killed
by frost, and regeneration will occur from the woody base
when temperatures rise. Leucaena does not naturally persist
under intense temperate winters, as those generally north
of 35°N latitude.

Leucaena is notably drought hardy, tolerating annual
rainfalls as low as 600 mm (5). It withstands annual rainfalls of 360 mm near Campina Grande in Northeast Brazil, and
survived over 10 years under 700 mm annual rainfall at Gayndah,
Queensland, while other palatable legumes died. It has become
naturalized as a major forest legume of sub-arid tropical
lowlands, notably in the Pacific Basin and Southeast Asia
(Figure 6). Its drought resistance is due in part to its
strong, deeply penetrating root system. It may be aided by
the ability of leucaena's tiny leaflets to intercept moist
air causing fog drip. During the intense 6 to 9 months' dry

Figure 6.   Leucaena dominates forests in Hawaii with annual
rainfalls limited below 400 mm.

season of many tropical areas, leucaena is often the only
legume producing some green foliage. It thrives under
irrigation in hot dry conditions as those of western Mexico
and Western Australia.

Leucaena does not tolerate soil conditions where flood-
ing or waterlogging persists for extended periods of a month
or so. It gives its highest yields in fertile tropical soils
which are deep, friable and moist, and neutral to alkaline in
reaction. It is an aggressive competitor in one of its
main centers of origin, the Mexican Yucatan Peninsula, which
is an uplifted limestone plateau with soils of pH 7 and above.
In spite of salt spray it grows on coastal coralline forma-
tions down to sea level in the Hawaiian, Philippine and other
tropical islands (Figure 6). Leucaena does not compete
favorably with natural vegetation above 1500 m in the tropics,
an apparent result of its lower growth rates at low
temperatures. It can however be cultivated at high elevations, and several related L. spp. are expected to thrive
at these elevations (notably, L. esculenta and L. diversifolia).

Yields of leucaena are highly light- and temperature-
dependent, although little studied physiologically. Forage
yields in Hawaii follow temperature and radiation values
very closely from the sunny summer months (avg. 26°C and
550 langleys) to winter months (avg. 21°C and 350 langleys).
Daily dry matter increments under these conditions ranged
from 13.8 to 26.9 kg/ha/day in 7 harvests of the Salvador
type (8). Harvest frequency must be varied during the year
to maximize forage quality, and ranged generally from 60
to 120 days and averaged about 90 days in the studies cited.
In Venezuelan midlands at 1450 m (avg. 21°C), harvest
intervals averaged 150 days for maximum yield of a Hawaiian-
type leucaena (16).

Leucaena does grow in some acid tropical soils, if the
seed is lime-pelleted and inoculated with adapted specific
Rhizobium strains, with attention given to soil nutrient
deficiencies (17). In most acid tropical soils the commonest
nutrient deficiencies for legumes include P, S, Ca, Mo, Zn
and sometimes Cu. CIAT has a breeding program on the
adaptation of Leucaena to the very acid (pH 4-4.5), high
aluminum soils of the South American tropics.

Establishment and Inoculation

Leucaena is commonly reproduced from seeds, since
vegetative propagation has not been sufficiently effective
or economic for wide-scale use. The seeds are small, brown

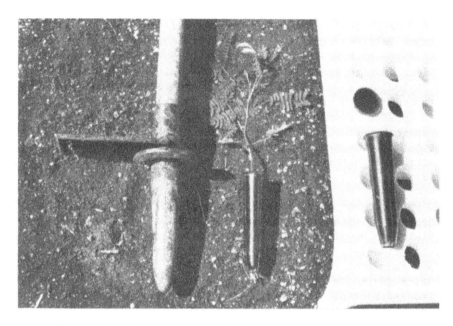

Figure 7.   Forest tree dibble tubes are well suited
to growth of leucaena seedlings and
transplanting at age of 3 months, as
shown.

and covered with an impervious coat that must be scarified to permit water-uptake and germination. Adequately dried and free of insects, the seeds apparently can be stored for years with little loss of viability. Scarification is most effective by nicking individual seeds with a knife or scissors. Larger seedlots can be scarified effectively by treatments with hot water (80°C for 3 minutes), with conc. sulfuric acid for 10 minutes, or with commercial carborundum-faced drum scarifiers. Seed weights average 20,000/kg for large-seeded Salvador or "giant" types of leucaena, ranging to 35,000/kg for smaller seeded varieties.

Leucaenas can be seeded directly, although establishment is slow, with seedlings attaining about 30 cm growth in 6 weeks. Planting densities of 75,000 to 100,000 trees per hectare have been recommended for forage management of leucaena, with rows spaced 50 to 75 cm. The seeds can be planted by ordinary grain drills, or dropped by hand or jabber into shallow trenches or holes. Weed control is critical for good establishment in most tropical areas. The herbicides alachor and nitrofen have proven to be the best preemergence treatments in Hawaii, at concentrations typical of those for soybeans and lettuce. Triazine herbicides like simazine can also be used for transplanting, if applied with caution on some soils. Land preparation should include consideration of leucaena's pH, P and rhizobial requirements.

Leucaena is inoculated by rhizobial strains that appear to be widely distributed although not universal in tropical soils. The bacteria nodulate other Mimosaceaous legumes, evidently including Mimosa pudica and Acacia farnesiana. Strains differing in their effectiveness on acid soils have been identified and released in Australia (15).

Transplanting is the standard practice for forest establishment of leucaenas. Forestry plantings are normally at densities between 2,500 and 10,000/ha, with rows spaced 100 to 200 cm. The seedlings are strongly tap-rooted, branching sparsely and with sharply angled lateral roots. An ideal transplant pot is provided by the long, narrow plastic "dibble-tube" (Figure 7). These solid tubes are 12.5 cm long and 3 cm wide, tapering at the base to an opening 1.5 cm wide with longitudinal inner ribs that prevent root spiraling. The tubes are suspended in styrofoam containers that cause the emerging roots at base of the tube to air-prune naturally. After 10 to 12 weeks growth, leucaena seedlings can be pulled easily from the tubes with no root injury. When placed in holes of the same dimension in moist soil, establishment is rapid and survival very high (Fig. 5).

Figure 8.   Experimental plots of "protein bank" plantings of
leucaena showing one month regrowth (Photo courtesy
of A. B. Guevarra).

Many types of rooting medium have been used successfully for leucaena; soil-peatmoss-vermiculite mixtures are preferred in Hawaii. A thin layer of cracked rocks can minimize surface fungal problems, obviating fungicide treatments. Nodulation should be ensured by treating the seeds, or using well-inoculated soil in the seedling tubes. Leucaenas have been transplanted from cans and plastic containers of many types, as well as bare-root seedlings from seedling beds. Survival can be high under some conditions, but establishment rates appear to maximize with the forestry "dibble-tubes".

Forage

Leucaena is an important browse and pasture legume for cattle, horses, goats and sheep throughout the tropics, and leucaena leaf meal is often a valuable supplemental feed for swine, poultry, and rabbits (5). It is viewed with increasing interest in aquaculture, where pellets or leaf meal make suitable feed for several species of fish and shellfish. In the past decade an international market has developed for dried and pelleted leaf meal of leucaena, and extensive forage plantings have been made in several countries. Historically the leucaenas probably travelled around the world as feed for animals in sea-going vessels. Animal feeding and related research developed extensively in Australia (4, 11, 18) and Hawaii (8, 19, 20, 21, 22), and more recently in the Philippines, Fiji and other countries (23, 24, 25, 26, 27, 28).

Leucaena is the highest producer of forage protein among tropical legumes. It occurs in three principal management patterns as a feed and forage plant in the tropics. It is most commonly seen as a widely dispersed shrub in natural grasslands, often over-grazed in dry seasons. It is also planted in hedgerows in pastures, and often survives in fence rows as a relict of over-intensive grazing. Hedges are known to have been grazed continuously more than 40 years. Less commonly it has been planted and managed as a solid stand. Leucaena rewards intensive care with high yields of a high quality forage, while surviving impressive neglect and environmental stress.

Protein Banks

Leucaena establishment and management is probably optimized in solid plantings that can be viewed as protein banks for supplemental grazing or for high-nitrogen forage for animals (Figure 8). A high proportion of the tropics is covered with nitrogen-deficient grasslands, such as the Llanos and Cerrados of South America. The animal productivity

Table 3. Forage dry matter yields as T/ha/yr of leucaena varieties compared on three Hawaiian islands (8).

Variety	Type	Oahu	Kauai	Maui	Average
K8	Salvador	22.0	15.4	21.0	19.5
K67	Salvador	21.3	13.5	20.4	18.4
K28	Salvador	17.4	14.1	20.1	17.2
K62	Salv./Peru	16.6	11.9	20.4	16.3
K4	Peru	9.4	6.6	20.4	12.1
K63	Hawaiian	10.0	5.2	16.3	10.5

Table 4. Forage and protein production of Hawaiian-type leucaena as influenced by frequency of cutting (22).

Days to Harvest	Dry Forage Yield	Protein	Protein Yield
	-T/ha/yr-	-%-	-T/ha/yr-
60	16.0	22.0	3.52
90	20.6	16.7	3.45
120	18.8	14.6	2.74

Table 5. Yields of leaf meal as dry matter, and protein of Hawaiian-type leucaena as influenced by frequency of harvest (19).

Days to Harvest	Dry Leaf Yield	Leafy Proportion	Protein	Protein yield
	-T/ha/yr-	-%-	-%-	-T/ha/yr-
75	10.8	79	28.1	3.03
100	12.3	69	24.8	3.05
125	14.0	58	25.9	3.63

of these vast areas could be increased with closely planted
protein banks of leucaena, rotationally grazed in
conjunction with the native grasses, e.g., with 10 ha of
leucaena per 100 ha of pasture. Fencing for controlled
access optimized management of the system to reduce liveweight losses during the dry season, and increase significantly the overall animal productivity. Establishment of protein
banks could be achieved at the beginning of rainy seasons by
seeding or by transplanting nursery-grown plants into fertilized furrows in the native savannah.

Yields of edible forage from experimental leucaena
plantings under monoculture maximize around 20 T/ha/yr as
dry matter. Great variability in yield and quality of forage
occurs among varieties in different ecosystems, and from
different frequencies and methods of harvest. The Salvador-
and Peru-type varieties normally outyield those of the
Hawaiian-type when properly managed, although increases in
forage yield are accompanied by greater woodiness (8, 18, 19,
27). Outstanding varieties in yield trials in Hawaii have
generally been of the Salvador type (Table 3). Yield deter-
minations in the Virgin Islands (USA) also confirmed the
superiority of the Salvador-types, with yields up to 20.5
T/ha/yr of edible dry matter over a 5-year harvest period
(27). In a subsequent Caribbean study of shorter duration
the dry matter yields of two Salvador strains averaged 17.5
T/ha/yr and of two Peru-type varieties averaged 14.0 T/ha/yr,
both well in excess of the local "Hawaiian-type" (27). Early
Australian studies simulating animal grazing over a 9-month
period also stressed varietal differences. Yields adjusted
to an annual basis ranged up to 16.8 T/ha for a Peru-type
variety in Australia (18).

The frequency of harvest or grazing must be adjusted to
the varietal characteristics to maximize protein yields.
The shrubby Hawaiian-type cultivars branch profusely and
flower rapidly following cutting (50 to 70 days), and must be
harvested or grazed frequently. Early studies in Hawaii's
lowlands (21, 22) suggested an optimum 70 to 90 day harvest
cycle for Hawaiian-type varieties (Table 4). Forage yields
were presented as a composite of edible leaf and fine stem
portions, with only coarse wood removed. A 60-day harvest
cycle resulted in lower forage yields but higher quality and
maximal protein yields (22), while a 120-day cycle led to
reduced forage quality due to high woodiness and seediness.

Similar studies by Guevarra (19) and by Hutton and
Beattie (11) focussed on the edible leafy portion obtained by
stripping branches, as if by a grazing animal (Table 5). The
leafy portion yields increased with time between harvests, as

did protein yields. Increasing the days between harvests had
the adverse effects of reducing the proportion of leaf to
stem, and reducing protein percentages. The arboreal Peru-
and Salvador-types appear to maximize yields of forage when
harvested less frequently, and when cut higher above the
ground. The ultimate choice of variety and harvest interval
thus can affect greatly leucaena's yield and forage quality.
These decisions will further be affected by the economics
of harvest, whether by animal, hand, or machine.

Machine harvest and post-harvest management of leucaena
were studied by Kinch and Ripperton (21), who engineered
some modifications of forage harvesting and drying equipment to handle leucaena. A drum-type forage dehydrator
dried the forage in a few minutes, and coarse stem separation
was achieved by modifying the stone separator on the drum.
Drying to 15% moisture was adequate to control mold.
Modern types of forage harvesters permit completly mechanized
harvesting of leucaena. Cubing or pelleting of the forage
should also become common in the future.

Solid stands of leucaena produce comparable yields over
a wide range of plant populations (19). Yields probably
optimize around 75,000 to 100,000 plants per ha (Figure 8).
Following a period of 4 to 8 months establishment, the plants
are cut down to between 10 and 50 cm in height to force
development of a highly branched crown. Under cooler
temperatures growth rates are slowed and the growth habit
changed. Herrera (16) reported maximal yields of about 10
T/ha/yr (dry matter) or Hawaiian-type leucaenas at 1450 m
elevations in Venezuela, when plants were grown to a height
of 150 cm and cut down to height of 75 cm. At this elevation
cutting to the height of only 10 cm was too drastic, greatly
reducing yields (16). The cutting height must also be adjusted upward for the arboreal Peru- and Salvador-type
varieties, whose potential forage yields are not realized
when harvests are cut to heights of 10 cm or less (19). In
small farm management systems, leucaena is best treated as a
source of cut-and-carry foliage. Careful pruning of the
shrub in solid plantings can lead to high quality forage,
but the yields per man hour are increased by cutting the
plants close to the ground, rather than careful hedge
trimming (Table 6).

Hedgerows

Leucaena is a shrubby legume that can be managed
efficiently in a hedgerow or fencerow, as a source of browse
forage or cut-and-carry green feed. Hedgerow management is
especially adapted to animal grazing, with the interspaces

Table 6. Forage yields of leucaena from hedgerow harvests at varying heights of rows spaced 1 m apart (22).

Height of Clipping	Forage Dry Matter Yields	Yield per Man Hour	Leafy Portion
-cm-	-T/ha/yr-	-kg-	-%-
10	14.2	51.5	36
50	12.1	36.7	41
100	11.2	26.0	46

Table 7. Forage yields of leucaena from hedgerow harvests at varying heights of rows spaced 3m apart (26).

Height of Clipping	Forage Dry Matter Yields	Yield per Man Hour
-cm-	-T/ha/yr-	-kg-
10	9.3	20.8
150	14.3	5.2
300	22.8	4.5

planted to grasses like guinea (<u>Panicum</u> <u>maximum</u>), pangola
(<u>Digitaria</u> <u>decumbens</u>), brachiaria (<u>B.</u> <u>decumbens</u>) or star
grass (<u>Cynodon</u> <u>dacylon</u>). Hedgerows can be planted in grass
pastures with 2 to 4 m between leucaena rows, a practice
becoming widespread in Mexico. The legume provides important
protein supplement to the animals and some nitrogen to the
associated grass, while often extending grazing periods late
into the dry season. The principal problems occur at times
of establishment and during over-grazing periods, when the
legume may fail to survive. The leucaenas also may grow into
small trees above the heads of animals during periods of
reduced grazing pressure, and may require periodic pruning
with machete or heavy field equipment.

Assessing forage yields of hedgerows, as obtained by
grazing cattle, is experimentally difficult. The ultimate
assessment must be in animal gains. Comparatively few such
trials have been conducted, most of them summarized by Hill
(5). They generally indicate improved meat or milk gains
with the added legume. Hedgerow yields have been assessed in
the Philippines with rows spaced 3 m apart and plants 5 cm
apart, harvesting 4 times annually to simulate pasture grazing
(26). Forage yields increased greatly with increasing height
of hedge in these studies, as plants expanded to intercept
more light (Table 7). Protein quality was decidedly superior
when the taller plants were trimmed as cut-and-carry feed.
However, yield per man hour maximized at the lowest cutting
height (cf. Table 6). All heights of cutting led to highly
economic returns. In related studies (26) it was concluded
that yields maximized when grazing was simulated to leave at
least 25% of the green leaves on the plant, as seen below:

Percent leaf harvested	Yield in T/ha/yr dry matter
100	17.8
75	20.5
50	15.6
25	12.8

Hedgerow data illustrate the great range of variation
encountered as variety and management practices of leucaena
are changed, and urge experimentation adapted to local
conditions. Pen-feeding and containment of browsing animals
makes hedgerow harvest most effective. Controlled feeding at
periodic intervals can be maintained so as to optimize yields
but minimize plant injury, then allowing recovery periods of
6 or more weeks for vigorous regrowth of the legume. Management of dairy cattle plantings must take into consideration

damage to udders by the cut plants, and the use of rubber-
tired equipment similarly presents spacing problems for
leucaena-grass pastures.

Feed

Leaf Meal

There has been an upsurge of interest since 1975 in
international marketing of leucaena leaf meal and pellets for
animal rations, as it is an excellent substitute for alfalfa
meal (21). Its use may be expected to expand as production
increases, and as alternate uses expand, e.g. in aquaculture.
High yields of harvested leaf meal can be obtained from
closely planted and well managed leucaena plantings, using
mechanized harvesting and artificial drying. Kinch and
Ripperton (21) reported yields from Hawaiian-type varieties
of 17.7 T/ha/yr of forage dry matter from 22 harvests on an
average 80-day cycle in Hawaii. Leucaena branches can also
be cut by hand and sun dried on a hard surface, the shed
leaflets then being swept up and bagged (7). The cut
material can be dried over wire netting which allows the
leaflets to fall through onto a collecting surface.

Whether harvested from hedgerows or protein banks,
leucaena leaf meal is easily obtained free of woody matter
and is of very high nutritive value. Mechanical harvests
of Kinch and Ripperton resulted in about 50% leaf meal (dry
matter basis), with the remainder equally divided among
coarse and fine stems (Table 8). As plants become older,
these ratios naturally change. When Hawaiian variety K341
was harvested 3 times vs 5 times annually the forage fraction
(leaf and fine stem) was reduced from 81% to 60% in the
studies of Guevarra (19).

Composition and Nutritive Value

The moisture content of the fresh forage fraction from
leucaena is normally about 75%, and drops very rapidly with
time after harvest. Field wilting may be feasible with
leucaena as it is with alfalfa hay, but leaflet loss begins
in 4 days. Data on the composition of leucaena forage and
leaf meal are summarized in Table 9 (5, 7, 16, 20, 21). The
high contents of crude protein and N-free extract emphasize
leucaena's value as animal feed. Fiber and ash values
range widely depending on the method and time of harvest.
Leaf meal fiber and ash contents are normally about 15% and
10% respectively (dry matter basis). Leucaena leaf meal is
relatively rich in potassium and calcium, with phosphorus
levels comparable to alfalfa. The meal can be viewed as a

Table 8. Composition of dehydrated leucaena following 8 harvests on 80-day cycle (21).

Component	Proportion	Protein	Carotene
	-%-	-%-	-ppm-
Leaf meal	50	30.0	434
Fine stems	27	13.3	81
Coarse stems	23	6.7	--

Table 9. Composition of leucaena as percentages of dry matter (7, 16, 22).

Component	In Forage*	In Leaf Meal
Fiber	32-38	18-20
Ash	6-7	10-11
Total N	3.0-3.5	4.0-4.3
Crude Protein	14-22	24-33
N-free extract	34-51	40-44
Phosphorus	0.27	0.23
Potassium	1.4	--
Calcium	0.8	--
Sulfur	0.14	--

*Forage includes fine stems normally consumed by grazing ruminant animals.

very good source of dietary minerals, and also as a high-
nitrogen green manure or fertilizer. The foliage appears to
be comparatively deficient in sodium (.02%) and high in
tannins (1.0%). No reliable evidence has been provided to
indicate that leucaena accumulates selenium; early reports
of animal symptoms ascribed to selenium were in error (5, 10).

Protein levels of leucaena leaf meal generally exceed
28%, averaging about 15% in green stems and 7% in coarse
stems (Tables 4, 5, 7, 9). Crude protein values are based on
N levels, which include a relatively high amount of N (0.5%)
from the alkaloid mimosine, to be discussed below. Average
protein values for whole forage (including fine stems)
increase as the frequency of harvest increases (Tables 4, 5).
The values were little affected by variations in plant spacing studies by Guevarra(17). In nature, Leucaena rarely shows
N-deficiency symptoms, such as in waterlogged soils, where it
is not nodulated, and during plant senescence.

The proteins of leucaena forage are of high nutritional
quality, and compare favorably with those of alfalfa in
amino acid fractions (Table 10). The S-containing amino
acids cysteine and methionine appear to be the first-limiting,
followed by tryptophane, based on non-ruminant requirements (5).
In ruminant animals the microbial protein adequately supplements these amino acids, unless sulfur is limiting in animal
diets.

Leucaena is a rich source of carotene and vitamins
(Table 8). The carotene or provitamin A content of leucaena
leaf meal is among the highest for plants (2-3 times that of
alfalfa), and the meal is widely respected as an additive to
poultry feeds to deepen the yellow color of egg yolks. The
yellowing effect extends to the fats of poultry and cattle
fed heavily on leucaena, an effect that is considered more or
less desirable depending on food habits. Leucaena foliage
also provides high levels of Vitamin K and riboflavin, about
double those of alfalfa.

Toxic Effects

In common with most legumes, leucaena produces some toxic
effects in animals when ingested at very high levels (see
review (6) by Oakes). Most conspicuous of these is the loss
of hair in humans and non-ruminant animals, and the reduction
of egg production in poultry. All toxic effects have been
shown to be reversible. Animal death has not been attributed
to leucaena feeding, unlike the effects of cyanogens, selenium
and saponigens (bloat -causing agents) in temperate legumes
like alfalfa and clover. However, fetal resorption can occur

Table 10. Amino acids as mg per g of N in
leucaena leaf meal compared with alfalfa
(5, 7, 40).

Amino Acid	Leucaena	Alfalfa
Arginine	294–349	357
Cystine	42–88	77
Histidine	112–125	139
Isoleucine	451–653	290
Lysine	313–349	368
Methionine	83–100	96
Phenylalanine	250–294	307
Tyrosine	232–263	232
Valine	255–338	356

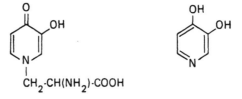

Mimosine   3,4-dihydroxypyridine

Figure 9. Chemical formulas for the amino acid
mimosine and its derivative, DHP
(dihydroxypyridine).

in pregnant swine and rats fed on leucaena. There is no
evidence of carcinogenic or mutagenic effects from leucaena
ingestion.

Horses that are ill-fed and allowed to browse heavily on
leucaena produce shaggy manes and tails. Cattle fed
excessively on leucaena, with over 40% of their dry weight
intake as the legume, have hair loss, poor growth and
excessive salivation. These symptoms appear to relate to
thyroid malfunction similar to goiter (37). Dairy cattle
fed high levels of fresh leucaena produce milk that is noti-
ceably off-flavor, an effect that is removed by pasteurization. Poultry fed more than 15% leucaena (dry weight
basis) show reduced egg production. Fetal resorption,
reduced birth weights and teratogeny have been recorded in
pigs (29). Reduced conception or reproduction has been
experimentally induced in rats fed on leucaena.

Mimosine (Figure 9) accounts for many but not all of
leucaena's toxic effects. Mimosine is an amino acid, (β -
N - (3-hydroxy-4-pyridone) - α aminopropionic acid (30, 31,
32). The amino acid occurs in very high levels in leaves
and seeds of leucaena, averaging about 4% dry weight in the
species L. leucocephala (33). Addition of the purified
chemical mimosine at high levels to rat diets causes atypical
oestrous cycles, infertility, and hair loss (4, 33), while
feeding of leucaena leaves in excess of 30% of diet reduces
fertility (34). Mimosine is readily precipitated in
solution by ferrous sulfate, and this is used as the basis
for its chemical assay (35, 36). It is reduced considerably
by soaking leaf meal overnight in water, or by heat treatments (23, 36). Several studies indicate that the addition
of 1% or less of iron sulfate during leucaena feeding can
eliminate or reduce substantially the mimosine toxicity in
pigs and poultry (25, 29, 36).

In ruminant animals, mimosine can be rapidly coverted
by bacteria in the rumen to DHP, dihydroxypyridine (Figure
9). Leucaena toxicity symptoms in such animals appear to
be due largely to the effects of DHP, an acknowledged
goitrogen (37). Neither mimosine nor DHP is found in blood,
milk, or meat, but DHP appears rapidly in the urine of
animals after feeding on leucaena. Sheep fed heavily on
leucaena evidently convert mimosine more slowly in the rumen,
permitting mimosine levels in the wool follicles that are high
enough to cause shedding of fleece.

Mimosine occurs in all Leucaena spp. tested, ranging from
2 to 6% on a dry matter basis in leaf meal. Wide genetic
variations occur in mimosine levels, with particularly low

levels in species like L. pulverulenta and varieties such as
those from Colombia, S.A. (13, 38). Advanced generations from
crosses with these lines show promise in reducing, but not
eliminating, the mimosine. Segregations were interpreted as
polygenic and uncorrelated with vigor or protein levels, but
varieties with the lowest mimosine presently available in
nature have been poor in vegetative vigor (13).

Animal Performance

Cattle. Pastures incorporating leucaena have demonstrated some of the highest carrying capacities for beef cattle in
the tropics, between 5 and 7 animals per hectare. Animals
rotationally grazing these pastures have produced liveweight
gains up to 1 kg per day. In one study in Queensland,
Australia, steers and calves averaged 0.93 and 0.88 kg/day,
respectively, when fed over a 7-month period on pastures of
leucaena and setaria by R. A. Jones (7). In Fiji the
addition of leucaena equal to 20% of native grass (Dicanthium
sp.) increased annual liveweight gains from 110 kg/ha on
the pure grass to 270 kg/ha in a 3.5 year study by
Partridge and Ranacou (28). The liveweight gains on leucaena-
supplemented pastures generally range between 0.5 and 0.7 kg/
day, over five-fold that for typical pastures of the tropics
(5, 7, 28). In vitro digestibility of leucaena is about 60%,
somewhat below the average for forage legumes, an effect that
may be due to its tannin content. Successful fattening and
marketing of steers can be achieved on balanced pastures
of leucaena and grass following as little as 3 months grazing
in Australia. In the drier tropics annual stocking rates and
liveweight gains will be reduced, but leucaena's drought
tolerance often insures the prevention of liveweight losses
after shallowrooted grasses dry down. Even on very high
leucaena intakes, liveweight gains of 0.55 kg/day occurred in
early studies of Henke et al (20) in Hawaii. Feeding of
pregnant cows on pure leucaena is neither sensible nor
economic, and is known to lead to a reversible goiter in
calves. In other monogastric animals, heavy feeding of
pregnant animals may lead to fetal resorption.

Leucaena is a valuable supplement to grass pastures for
dairy cattle with controlled intake of the legume. Milk
yields are reported up to 6000 1/ha annually on such pastures.
Henke (20) fed leucaena as the sole roughage in 3-year trials
of dairy cattle that consumed an average of 21 kg/day (dry weight)
of leucaena. Milk gains of 8% were observed versus
the controls fed on Napiergrass roughage, and off flavors
were not reported. The value of leucaena for milk production
has been realized only recently by Australian dairy farmers;
pasteurization removed off-flavors found in fresh milk and

cheese. These flavors can be avoided by eliminating leucaena
from the diet at least two hours before milking . They
occurred at levels of intake generally in excess of 10% dry
matter in the diet, and would probably be less offensive in
most of the tropics.

Poultry. Leucaena is a valuable addition to poultry
rations when properly used, and stimulates growth and egg
production (39). Egg yolks develop a desirable deep yellow
color from the high carotene content of the meal, and
leucaena achieves this effect at less than half the intake
level of alfalfa. Controlled feeding has been reported to
increase egg hatchability, due to high riboflavin and
Vitamin K contents of leucaena. A level of 5 to 6% (dry
weight) is considered safe for broilers or layers with all
types of leucaena meal. Egg production and broiler weight
gains are reduced, however, when leucaena meal is fed at high
levels, notably in excess of 10% dry matter (23, 24, 40).
The effects are similar to those obtained with alfalfa and
other legumes, and evidently relate as much to protein
digestibility as they do to toxicity of leucaena (24, 40).
Treatments of leaf meal that reduce mimosine levels greatly,
such as overnight soaking in water or hot water and steam
treatments, appear to improve the feed value of leucaena (23,
37). Similarly, it is reported that ferrous sulfate
supplements of 0.3% improve efficiency of poultry rations containing leucaena (25).

Sheep and Goats. Ruminant animals can be adapted in
short periods to detoxify the mimosine of leucaena and accept
proportions of leucaena in the diet comparable to that for
cattle. Goats will browse leucaena heavily without hair loss,
and it can be valuable supplemental feed for them. It is
dubious that off-flavors of goat milk and cheese, if they
occur, would be detected; in any event, none have been
reported.

For sheep, the risk of fleece-shedding by unadapted
animals is presently too great to encourage indiscriminate
leucaena feeding (41). Animals fed excessively high amounts
of leucaena for short periods can actually be sheared by
hand. Commercial chemical-shearing is currently being
studied with other compounds (7). Heavy feeding of pregnant
sheep on leucaena can reduce birth weight of lambs, and cause
goiterogenic symptoms (5).

Swine. Leucaena is coming into increasing use as a high-
protein supplement for growing and fattening pigs. Up to 10%
leaf meal (dry weight) has proven a satisfactory supplement
and normally does not cause side-effects. Weight gains

Figure 10. Leucaena K8 windbreak planting 6 years
of age, with auracarias of same age.

remain high on feeds supplemented with leucaena up to 30% of diet (20% is recommended), when mineral supplements including ferrous sulfate are provided (25). For breeding gilts, however, it is recommended that leucaena meal be withheld during breeding to avoid increased fetal resorption (29). Alternatively, the leucaena meal must be supplemented with ferrous sulfate (1% recommended) throughout the critical period of conception and early development.

     Fish and Miscellaneous. Leucaena has been fed successfully in very limited studies to several species of tropical fish, and might become an important high-protein fish meal. Tilapia showed normal gains and fertility when fed leucaena leaf meal, placed inside feeding rings, at levels of 2% and 4% of the fish body weight daily (42). Leucaena trees around aquaculture ponds can also provide desirable shade and wind protection, and the dropped leaflets add nitrogen to the ponds. Rabbits relish leucaena, although it must be fed under controlled conditions to less than 10% of diet (dry weight) to avoid hair loss. Water buffaloes can consume leucaena in the amounts recommended for cattle. Horses and donkeys throughout the tropics can be seen with shaggy manes and shortened tails, a reflection of their penchant for leucaena. They should normally be tethered or fed so as to control intake to less than 10% of dry matter.

Energy and Wood

     At least half of the world's harvested timber serves as fuel, the majority of this in developing countries. Many scientists feel that the approaching fuelwood crises of these countries will create an impact as severe as food crises of the past, since tropical forests and fuelwood resources are diminishing at an alarming rate. Leucaena is among leguminous trees that could provide the ton or more of fuel required annually per person in the tropics with minimal use of land and maximal ecological effectiveness, as leucaena forests serve also for erosion control, watershed protection, etc. At maximal productivity, a single hectare of leucaena could provide fuel for at least ten families. A new conciousness is needed in most of the tropics, however, of the concept that fuel can no longer be gathered but must be grown.

Wood Fuel and Charcoal

     Leucaena has few peers as renewable resource of biomass energy (Figure 10). The "Hawaiian Giant" varieties of the Salvador type have produced wood yields among the highest on record (43). Trees in the Philippines representing 4 Hawaiian Giant varieties averaged 239 m³/ha/yr or 115.7 bone

Table 11.   Estimated annual wood yields and characters of leucaena giants in the Philippines (43).
DBH=diameter at breast height; bdt =bone-dry tons.

Age	Stand Density	DBH	Trunk, Clear Height	Yields of Trunk	
-yrs-	-plts/ha-	-cm-	-m-	$-m^3$/ha/yr-	-bdt/ha/yr-
2	2,500	9.2	4.6	26.5	14.0
2.5	44,000	8.9	2.4	239.3	115.7
7	3,000	14.9	7.4	45.6	25.5

Table 12.   Combustion values of Hawaiian-type
leucaena in Indonesia (3) as compared with softwoods and fuel oil.

Fuel	Combustion Value	Moisture	Ash
	-cals/kg-	-%-	-%-
Leucaena wood	3,895*	10.9	1.6
Leucaena charcoal	7,250	--	1.0
Softwoods, air dry	2,700	30.0	1.2
Fuel oil	10,000	--	--

*Philippine samples of Salvador type (7, 43)
 ranged from 4,170 to 4,680 cal/kg (or 7,460 to
 8,380 btu/lb), with 0.8% ash.

dry tons/ha/yr (52.6 bone dry tons/acre/yr), when harvested
at 2.5 years of age in a closely spaced plantation (Table 11).
The trees averaged 9 cm in diameter and 2.4 m clear height
of wood (43). At very low spacings under 3000/ha trees
doubled in height with little increase in diameter, but
wood yields were greatly reduced. A point of reference for
these data would be commercial harvests of eucalypts in
Hawaii that are economic averaging 2.5 bone dry tons/acre/yr.
Under improved conditions, however, species like Eucalyptus
deglupta, Albizia falcataria and Gmelina arborea produce
annual growth increments of 10.8 to 11.9 bone dry tons/A/yr
(43). Leucaena wood productivity ranges upward from 25 bone
dry tons/A/yr for a great variety of spacings and managements
in the Philippines and Hawaii, although critical comparative
studies of spacings, varieties and environments have only
recently been initiated.

These figures may be wasted on oil-rich countries and
those with an affection for space-age technology that seems
capable of harnessing the more exotic energy sources of the
earth, sea or sky. Vastly greater resources have been
allocated to research on geothermal, ocean thermal, hydraulic,
wind and solar energies than to biomass. However, biomass
energy of trees like leucaena may be the cheapest, most
environmentally sound, renewable source in much of the world.
Technology is sorely needed on production, harvest, utilization, gasification, charcoal and alternate uses of leucaena.
Leucaena trees can be harvested repeatedly; e.g. more than
60 years' continuous harvest in a large area near Los Banos,
Philippines. These plantations do not require the expensive
nitrogen fertilization that is essential to maximize biomass
productivity of most other tropical forest trees.

Leucaena wood is dense with minimal bark, and burns well
and cleanly (Table 12). Few combustion values are reported
for leucaena air dry wood. These range from 3800 to 4700
cal/kg*, substantially above soft woods (Table 12). The
wood burns with little smoke and low ash, less than 1% in
the giants (43) It makes a superb charcoal, and is widely
favored for this use in Southeast Asia. Several thousand
hectares of giant K8 are now coming into production in the
Philippines as a source of commercial activated charcoal.
Leucaena charcoal has a heating value over 70% of that of
fuel oil, and can be recovered from simple pits or retorts,
a potentially important small industry near plantations.
Charcoal has major advantages of ease in shipment and

_____

*One bone dry metric ton of leucaena thus easily exceeds
in combustion energy a barrel of oil.

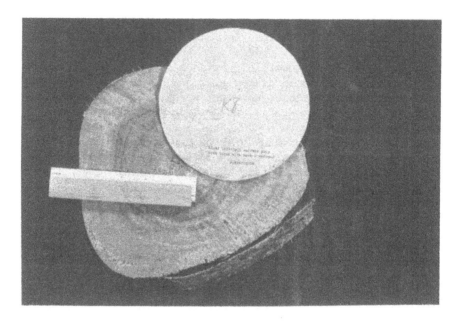

Figure 11.   Cross-section of 7-year old leucaena
giant together with 6" ruler, parquet
floor strip and paper produced in the
Philippines.

Table 13.   Fiber characteristics of Salvador-
type leucaena, averaging values for trunks and
branches (43).

Fiber characteristic	Description
Length (L)	1.14 mm
Width (W)	0.026 mm
Slenderness ratio (L/W)	44
Lumen width (l)	0.016 mm
Flexibility ratio (l/D x 100)	62
Cell wall thickness (w)	0.005 mm
Runkel ratio (2w/l)	0.63
Runkel group	I
Mustelph Value and Group	64%, III

storage, and clean burning. Industrial estimates by the Mabuhay Vinyl Corporation in Mandanao, the largest producer of leucaena charcoal in the world, target optimal charcoal production at 30 T/ha/yr, equivalent to 300 m^3 of wood, when trees are harvested on a 4-year rotation cycle.

Pulp and Paper

Although studies of pulp and paper from leucaena are limited, they suggest that leucaena wood has great commercial potential especially as processed for paper products and particleboard (Figure 11). It is a medium hardwood with density similar to koa or ash. Densities ranging from 0.5 to 0.7 have been reported from trees of differing ages (1.5 to 8 years), different varieties and growing conditions. Salvador-type trees of 2 to 7 years in age ranged from 0.48 to 0.56 in density (43). Moisture contents of 60 to 70% characterized the mature wood, and bark accounted for less than 8% of the bone dry weight (Figure 11).

The fiber characteristics of leucaena are typical of hardwoods, and are in an acceptable range for paper (Table 13). The fibers are relatively short, but have a good slenderness ratio for papermaking. Similarly, the Runkel ratio that relates cell wall thickness to diameter of cell cavity (lumen) is less than 1.0, ideal for pulpwoods.

Pulping characteristics of leucaena are like those of other tropical hardwoods. Proximate analyses proved similar to the average for 95 Philippine hardwoods (43), with 71% holocellulose, 15% pentosans and 24% lignin. The relatively high holocellulose value (71%, vs. 64% average for hardwoods) is interpreted as an indication of good pulp yields. The silica and pitch contents (43)were very low (solubility in alcohol-benzene was low at 2%). Two unpublished studies of giant leucaena, taken under different kraft cooking conditions, indicated pulp yields of 44.0 and 51.3% (based on oven dry weight), the latter a very respectable figure. Due to the high wood density of leucaena, pulp digester yields were relatively high, and required larger amounts of alkali than softwoods like alder and pine.

Paper made from leucaena as kraft handsheets (Figure 11) show characteristics overlapping that of of fast-growing pulpwoods (7,43). The paper had fair tearing strength but low folding endurance and tensile strength. It was considered a good tissue pulp for its opacity and printability, and would be used principally to blend with long-fiber woods in the production of printing paper. Other uses of leucaena pulp could be in the production of rayon and cellophane.

Figure 13. Relict tree of L. diversifolia in high-land Salvador forest. Leucaenas provide important nutrition to tropical forests, but are rapidly being deforested.

Figure 12. Leucaena giant trees 5 years old in dense stand of 60,000/ha show self-pruning, erect habit and suppression of weeds and seedlings.

Posts, Lumber and Other Uses

The trunks of leucaena trees are of hardwood with increasing use as props for bananas, as fenceposts and as building posts. The common leucaenas are very irregular in growth and of little construction value. Tree conformity and bole shape of the giant leucaenas is directly related to the density of planting and management of trees. Isolated individual trees with no pruning tend to fork and branch at the base, with a semicircular crown whose diameter is about equal to crown height. As the density of planting increases, however, the trees rarely fork and become self-pruning, with relatively erect and unforked boles (Figure 12). At high plant densities of 10,000 to 60,000 per ha, the trees make good posts to prop bananas and to serve as girders and rafters in small building construction. The leucaena wood absorbs termite-resisting preservatives well, and can become large enough for small telephone and power poles; 8-year old Philippine trees ranged to 20 m in height and were erect and typhoon-tolerant.

Leucaena wood is close grained and is readily workable, lacking excessive gums, knots and variations in heart and sapwoods. The wood is yellowish white, with heartwood somewhat darker. The wood is expected to have increasing use as source of lumber or woodchips in production of particle-boards, fiberboards and plywood. The density of the wood makes it suitable for parquet flooring (Figure 11), and it can be turned fairly well as craft and sculpture wood.

Reforestation and Soil Restoration

Reforestation and Erosion Control

Unbridled deforestation occurs in the tropics today much as in the United States a century ago. Approximately half the native tropical forests have been lost within this century. Wood gathering, slash and burn farming, uncontrolled grazing, and irresponsible logging rapidly deprive the tropics of these natural forests. Soil erosion, floods, landslides and nutrient depletion follow. The major role of leguminous trees in constant soil renewal and nitrogen cycling (44) is perhaps to be discovered too late in the tropics, where woody legumes are a much more significant key to healthy forests than in temperate regions (Figure 13). Intensive year-round growth, rapid and continuing decay and leaching, and heat and rain conspire to deplete forests quickly of nitrogen when N-fixing species are removed. The poverty-stricken savannahs and shrubby climax vegetation that succeed are normally intolerant of drought, and are usually overgrazed and razed regularly by fires. Irreversible soil and nutrient loss and laterization

Figure 14.   Hawaiian-type leucaena serving as N-nurse and
             shade tree for coffee in the Cameroons, 11C0 m
             elevation.

can ensue. Leucaena is one of the best leguminous trees to
break this insidious cycle.

Leucaena can be planted in solid reforestation blocks,
but its greatest value would appear to be as an N-fixing
partner to other tree species. On timber lands, leucaena can
be a significant nurse or companion tree for eucalypt, teak,
or pine tree. In depleted forest lands, the judicious
interplanting of low-seeding giant varieties of leucaena could
help arrest the soil depletion process and provide a basis
for the evolution from grass back to forest. Many places in
Hawaii illustrate this natural recovery process since the
removal of intense grazing pressure four decades ago.
Leucaena originally invaded and dominated the impoverished
grasslands. As the tall forest species slowly returned,
leucaena was shaded out and ceased to dominate the forest.

Reforestation has been aided by the planting of leucaena
in Indonesia, in a program called "lamtoronisasi" (7). As in
the Philippines and New Guinea, leucaenas are planted on the
contours of steep slopes where they provide quick canopy and
organic matter for ground cover to further reduce losses to
soil erosion. The significance of leucaena as a soil-erosion
plant was lucidly and systematically outlined by Dijkman (3)
almost 30 years ago. It is more of a reflection of political
than of scientific judgement that Dijkman's compelling
treatise has been neglected so long. Reforestation in the
tropics appears to succeed only where sociopolitical muscle
reinforces the agroforestry expertise, since reforesting can
succeed only if indiscriminate land settlement, grazing,
logging and burning are confined. Dijkman (3) reviewed in
detail a half-century of Dutch, Indonesian and Central
American experience with reforestation and plantation renova-
tion, concluding that leucaena was "one of the most amazing
aid-plants known for tropical agriculture (p. 339).

Nurse and Shade Tree

Leucaena was dispersed through the tropics most widely
in the late 19th century as a nurse tree or shade tree for
plantation crops (Figure 14). Among these crops were coffee,
cacao, tea, quinine, rubber, mangosteen, pepper, citrus,
vanilla, and even coconut and oil palm. Extensive studies on
leucaena in plantations were made by Dutch scientists in Java
in the early 20th century, and are summarized by Dijkman (3).
The common leucaena grows into a small, many-branched tree
of 10 to 15 m in height when well cared for. Such trees
create a moderate shade and provide protection from wind and
storm in an orchard or plantation. The leucaena roots do not

compete with the plantation trees, as they penetrate deeply
with little or no lateral or superficial branching.

It is probable that leucaena's primary value in
plantations is often not for shade but for the high N leaflets that are dropped and act as fertilizer or green manure
for the plantation crop. Regular pruning of leaucaena side-
branches was practiced as early as 1910 in Indonesia.
Dijkman (3) reported the nitrogen return to soil from this
practice to be equivalent to one ton of ammonium sulfate
fertilizer per hectare per year, assuming 1000 trees/ha of
leucaena. The added contribution by dropped leaves of
phosphate, potash, calcium and other salts similarly leads
to soil enrichment of value in a plantation.

Shade is important for young trees of plantation crops
like cacao and coffee in some areas, and leucaena is considered an ideal shade tree that can be shaped easily to fit any
required space. The Hawaiian or common type is superior as
a plantation shade tree, although giant types were effective
at higher elevations in Indonesia (3). As shade trees the
giant types can become overpowering at lower elevations. The
species L. diversifolia (Figure 13) with its conspicuous
reddish flowers, also makes a good plantation shade tree.
Reseeding might be less of a problem with this species.

Leucaena giants should become superb companion and
nurse trees in the future for tropical energy and wood-
product forests. During early years of growth, forest trees
like the eucalypts intercept light inadequately, allowing
excessive weed growth and often severe erosion. Unless
fertilized continuously the eucalypts can take many years to
achieve harvestable size. Interplanted leucaenas can fix
nitrogen and provide it to the eucalypts, while controlling
erosion and weed growth during early years of plantation
development. It should then be possible to harvest the
leucaenas and eucalypts concurrently for many uses, as in
chipping. Leucaena has been planted as a nurse plant for the
slow-growing teak, and increases of 100% in the teak growth
reported in Indonesia (3). The ecological consequences of
incorporating leguminous trees into nonleguminous forests
clearly encourage further studies of integrated forest crop
management with leucaena.

Windbreak

The common leucaena occurs widely in the tropics as a
windbreak or protective hedge and fence, but is often cursed
for its seediness and rapid regrowth. Giant leucaenas appear
much more suitable to hedge management without such reseeding

problems. In the future vegetatively propagated, sterile
hybrids can be expected to replace the giants. Giants can
grow to 5 m in 6 months, 8 m in 2 years and 17 m in 6 years.
A single row of trees spaced 1 m apart is recommended, and
can be supplemented with a low-growing hedge like hibiscus,
panax or sugarcane (Figure 10). Unwanted seedling trees can
be killed by painting herbicides like picloram or 2-4,5 T
(where permitted) on cuts through the bark. Severe typhoons
rarely topple or break the trunk of the limber, deeply rooted
leucaenas. While the leaflets strip off in heavy winds or
prolonged drought, they are quickly replaced.

Firebreaks of leucaena have proven effective in stopping
or retarding the spread of grass fires. Leucaena remains
green late into the dry season, when millions of hectares in
the tropics are intentionally burned. Fireline strips of
20 m in width could provide inexpensive and low-maintenance
breaks.

Soil Restoration

Slash-and-burn farming is still practiced by 1/4 billion
farmers in the tropics. It relies on slow natural restoration
of nutrients to soil during long fallow periods between food
crops, periods that average about 15 years in the tropics.
This process can be accelerated greatly by supplementing
nitrogen, commonly the first-limiting element after tropical
crop production (44). Civilizations like those of the
Classic Maya (Figure 1) evidently may have mastered soil
husbandry to reduce the fallow periods between their corn
crops to 3 to 5 years in Northern Guatemala and Southeastern
Mexico (45). This area is a major center of diversity of
leucaena and several other mimosoids, and leucaenas are
typically seen near Maya ruins. Could leucaena have played
a major role in soil restoration for the highly effective
Mayan agriculturists? Preliminary efforts to integrate soil-
building by leucaena into shifting cultivation rotation in
the Philippines (7) and Papua New Guinea (46) have been very
promising and lend support to this thesis. In the majority
of denuded tropical soils, a very modest investment in
phosphate and lime would stimulate optimal leucaena growth
and assure rapid restoration of soil N. The deep rooting of
leucaena allows it access to water and nutrients that are
inaccessible to many other plants, and its leaf drop (Table
9) gradually returns these elements to build the topsoil (3).

Small-Farm Agroforestry

Leucaena is a tree ideally suited for use on small farms
in much of the tropics. It is easily established from seed,

Figure 15.   Leucaena intercropped with corn, at age 10 weeks;
             leucaena was clipped as green manure at time of
             corn planting (Photo courtesy of A. B. Guevarra).

versatile in use, and virtually trouble-free. It is most
commonly seen on the small farm as a hedge or individual tree
serving for shade, feed and food. The giant types are clearly being spread widely through Mesoamerica by being handed
from one farm to the next. Ignorance remains the major
barrier to rapid dissemination of improved varieties; as
examples, Latin Americans do not know leucaena as a fuel or
nurse tree, and Asians do not recognize its value as green
manure.

Green Manure

Leucaena leaf meal is a superb organic fertilizer,
applied fresh or dried, and is free for the harvesting in
much of the tropics. Annual nitrogen yields from regularly
harvested leucaena in the study of Guevarra (19) ranged
from 480 to 580 kg/ha (Table 5), and corn yields from the
applied meal were easily doubled even on fertile experiment
station soils (Figure 15). Control corn plots averaged 1.9
T/ha in one study; application of leaf meal equivalent to
150 kg of N per ha increased this yield to 6.6 T. Half the
leucaena was incorporated prior to planting and half applied
as a sidedressing. As with other green manures, the
efficiency in N utilization is probably less than 50%. However, most tropical corn farmers can afford to purchase no
added N, and for them efficiency arguments are academic. The
leucaena foliage of Guevarra's study also contained 44 kg/ha
of P and 187 kg/ha of K, as well as calcium, sulfur, magnesium and other nutrients (Table 9). Added organic matter
serves to improve soil tilth and aeration, water retention
and cation exchange capacity.

Faced with a general infatuation with inorganic
fertilizers and their obvious industrial support, leucaena
and other legumes have made little headway as organic
fertilizer in the tropics. Sadly, they may succeed only
where use is enforced or where industry capitalizes on their
sale. About 5 bags of leucaena leaves (4.3% N) can substitute for a bag of ammonium sulfate of the same weight.
Insect attack or deterioration do not occur if the meal is
adequately dried. One hectare of leucaena could serve a
large number (10-15) of corn farms, that now average 1.3
T/ha of corn (22 bushels) in the tropics, versus Guevarra's
6.6 T/ha.

As organic fertilizer, leucaena may be best planted as
wind- and erosion-controlling strips on the contour through
a small farm. Soil between these strips can be prepared by
incorporating leucaena into the rows or hills at a rate of 25
gm of dried meal (or 100 of fresh) per hill of corn, cassava

Figure 16.   Leucaena pods for sale in market in Chiapas,
              Mexico; the immature plump seeds are eaten fresh.

or bean. Side-dressing with an equivalent amount of leucaena
is recommended one month later (2). The use of composts
and organic fertilizers in the tropics has been curtailed in
part by the rapid deterioration of such materials when wet
and hot; cheap plastic bagging of sun-dried leucaena may be
an economic solution. It is noteworthy that the best rice
yields of the world occur in South Korea, where farms are
little larger than those cited in the tropics, but where
amounts of compost used are incredibly high (N-supplemented).
As fossil fuels and derived fertilizers become increasingly
expensive, legumes like leucaena should not have to beg for
consideration by agriculturists in the tropics.

Support Systems for Climbing Plants

A large number of tropical hroticultural plants rely
on support systems (trees in the native ecosystem) for
maximal yields, allowing them better light, aeration and
freedom from pests. Support systems have become increasing
expensive and labor-demanding, negating the benefits from
increased yields or crop value from their use. Preliminary
trials indicate that leucaena may be effectively managed as
support system for beans and lima beans, vanilla, black
pepper, passionfruit, yams (<u>Dioscorea</u> <u>spp</u>.), winged beans
and other tropical viny plants. Young leucaena trees can be
cut back to stimulate branching, then pruned as desired for
support. Alternatively, trees can be killed or felled;
yields of the favored yams of West Africa were maximized when
grown over such trees.

Food and Other Uses

Historically, the first use of leucaena may have been as
food. The state of Oaxaca, Mexico, derives its name from
"huaxin" (leucaena), locally known best as a food. The young
seeds are eaten raw and are considered a delicacy in Oaxaca,
Chiapas and neighboring states (Figure 16). Many households
in Southern Mexico grow a tree or two of leucaena and regularly harvest the young green pods. The seeds are also cooked
in soups, or into bean pastes that are dried and used as a
source of protein through the long dry season. The unpleasant
odor associated with such cooking is familiar throughout
Southern Mexico and wherever beads are strung from hot-water
softened seeds. Very young pods may also be served in
tortillas with chili sauce. Young leaves are also used in
soups in Southeast Asia. There is need to assemble information on leucaena's medicinal properties, for its reputation
in control of stomach disorders is widespread (6).

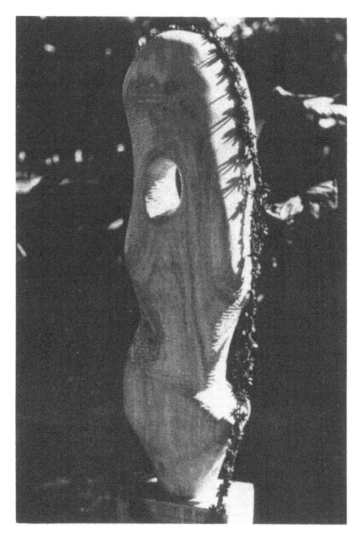

Figure 17. Leucaena's hard wood is gum-free and turns well
for sculpture; seeds are widely used in jewelry
(Statue courtesy of Francisco Verano).

The mimosine (Figure 9) of leucaena can cause hair loss in people as in animals. The effect was first recorded in Indonesia, and its prevention was related to the use of iron cookware. Iron causes the precipitation and removal of the offending mimosine (6). Overnight soaking would also be expected to remove most of the mimosine. No other hazardous side effects to man from ingestion have been demonstrated, but might be expected if sustained high intake occurred.

The seeds of leucaena are widely used throughout the world in ornamental and household items such as necklaces and hotpads, and the wood carves well for hardwood home products (Figure 17). The dry seeds can yield high quantities of gums of the galactomannan type (similar to gum arabic) of potential commercial significance (7). Natural dyes for wool and cotton goods are also extracted from leucaena pods and wood; red pods, notably from L. esculenta, produce red colors while the brown pods and wood give earth colors (7). Preliminary studies have suggested potential value of liquid protein concentrates and of hemagglutinins in leucaena.

Note

Journal Series No. 2346 from the Hawaii Agricultural Experiment Station, Dept. of Horticulture, University of Hawaii, Honolulu, HI 96822, and the Centro Internacional de Agricultura Tropical, Apdo. 6713, Cali, Colombia.

References and Notes

1. H.C.D. de Wit, Taxon 24, 349 (1975).
2. J.L. Brewbaker, in Produccion de Forrajes (FIRA, Banco de Mexico, 1976), pp. 13-27, 165-181.
3. M.J. Dijkam, Econ. Bot. 4, 337 (1950).
4. S.G. Gray, Tropical Grasslands 2, 19 (1968).
5. G.D. Hill, Herbage Abst. 41, 111 (1971).
6. A.H. Oakes, Advancing Frontiers of Plant Sci. (India) 20, 1 (1968).
7. National Academy of Sciences, Leucaena: Promising Forage and Tree Crop for the Tropics (Washington, D.C., 1977).
8. J.L. Brewbaker, D.L. Plucknett, V. Gonzalez, Hawaii Agric. Exp. Sta. Research Bull. 166, 22 (1972).
9. J.L. Brewbaker, Hawaii Agric. Exp. Sta. Misc. Publ. 125 (1975).
10. E.M. Hutton, S.G. Gray, Empire J. Experimental Agric. 27, 187 (1959).
11. E.M. Hutton, W.M. Beattie, Tropical Grasslands 10, 187 (1976).
12. S.G. Gray, Austral. J. Agric. Res. 18, 63 (1967).
13. V. Gonzalez, J.L. Brewbaker, D.E. Hamill, Crop Sci. 7, 140 (1967).
14. S.G. Gray, Austral. J. Agric. Res. 18, 71 (1967).
15. D.O. Norris, Austral. J. Exp. Agric. Animal Husb. 13, 98 (1973).
16. G. Herrera P., Agricultura Tropical 23, 34 (1967).
17. E.M. Hutton, C.S. Andrew, Austral. J. Exp. Agric. Animal Husb. 18, 81 (1978).
18. E.M. Hutton, I.A. Bonner, J. Austral. Inst. Agric. Sci. 26, 276 (1960).
19. A.B. Guevarra, A.S. Whitney, J.R. Thompson, Agron J. 71 (In press, 1979).
20. L.A. Henke, Proc. 8th Pacific Sci. Congr. 4B, 591 (1958).
21. D.M. Kinch, J.C. Ripperton, Hawaii Agric. Exp. Sta. Bull. 129 (1962).
22. M. Takahashi, J.C. Ripperton, Hawaii Agric. Exp. Sta. Bull. 100 (1948).
23. L.S. Castillo, F.B. Aglibut, A.L. Gespacio, L.S. Gloria, A.R. Gatapia, R.S. Resurreccion, Phil. Agric. 47, 393 (1964).
24. J.N. Hathcock, M.M. Labadan, Nutr. Rep. Int. 11, 63 (1975).
25. G. Malynicz, Papua New Guinea Agric. J. 25, 12 (1975).
26. R.C. Mendoza, T.P. Altamirano, E.Q. Javier, Herbage, crude protein and digestable dry matter yield of ipil-ipil in hedge rows (unpublished paper, Philipp. Soc. Animal Sci. 1975).

27. A.J. Oakes, O. Skov, <u>J</u>. <u>Agric</u>. <u>Univ</u>. <u>Puerto Rico</u> 51, 176 (1967).

28. I.J. Partridge, E. Ranacou, <u>Tropical Grasslands</u> 8, 197 (1974).

29. O. Wayman, I.I. Iwanaga, I. Hugh, <u>J</u>. <u>Animal Sci</u>. 30, 583 (1970).

30. R.L. Adams, J.L. Johnson, <u>J</u>. <u>Am</u>. <u>Chem</u>. <u>Soc</u>. 71, 705 (1949).

31. J.W. Hylin, <u>Phytochemistry</u> 3, 161 (1964).

32. R.K. Yoshida, Unpublished Ph.D. thesis, University of Minnesota (1944).

33. J.W. Hylin, I.J. Lichton, <u>Biochemical Pharmacology</u> 14, 1167 (1965).

34. B.M. Bindon, D.R. Lamond, <u>Proc</u>. <u>Austral</u>. <u>Soc</u>. <u>Anim</u>. <u>Prod</u>. 6, 109 (1966).

35. M.P. Hegarty, R.D. Court, P.M. Thorne, <u>Austral</u>. <u>J</u>. <u>Agric</u>. <u>Res</u>. 15, 168 (1964b).

36. H. Matsumoto, E.G. Smith, G.D. Sherman, <u>Arch</u>. <u>Biochem</u>. <u>and Biophys</u>. 33, 201 (1951).

37. M.P. Hegarty, R.D. Court, G.S. Christie, C.P. Lee, <u>Austral</u>. <u>Vet</u>. <u>J</u>. 52, 490 (1976).

38. J.L. Brewbaker, J.W. Hylin, <u>Crop Sci</u>. 5, 348 (1965).

39. M.M. Labadan, T.A. Abilay, A.S. Alejar, V.S. Pungtilan, <u>Philipp</u>. <u>Agric</u>. 53, 402 (1969).

40. J.P.F. D'Mello, D. Thomas, <u>Tropical Agric</u>. 55, 45 (1978).

41. M.P. Hegarty, P.G. Schinckel, R.D. Court, <u>Austral</u>. <u>J</u>. <u>Agric</u>. <u>Res</u>. 15, 153 (1964).

42. E.M. Cruz, I.L. Laudencia, Unpublished reports of Central Luzon State Univ. N. Ecija, Philippines (1976).

43. P.V. Bawagan, J.A. Semana, Unpublished report of Forest Products Res. and Ind. Dev. Comm., Los Banos, Philippines (1976).

44. R.A. Date, <u>Soil Biol</u>. <u>and Biochem</u>. 5, 5 (1973).

45. J.L. Brewbaker, <u>Econ</u>. <u>Bot</u>. 33 (In press, 1979).

46. R.L. Parfitt, <u>Harvest</u> (Papua New Guinea) 3, 63 (1976).

Milton Keynes UK
Ingram Content Group UK Ltd.
UKHW020024071024
449327UK00032B/2917

9 780367 171124